SpringerBriefs in Molecular Science

SpringerBriefs in Molecular Science present concise summaries of cutting-edge research and practical applications across a wide spectrum of fields centered around chemistry. Featuring compact volumes of 50 to 125 pages, the series covers a range of content from professional to academic. Typical topics might include:

- A timely report of state-of-the-art analytical techniques
- A bridge between new research results, as published in journal articles, and a contextual literature review
- A snapshot of a hot or emerging topic
- An in-depth case study
- A presentation of core concepts that students must understand in order to make independent contributions

Briefs allow authors to present their ideas and readers to absorb them with minimal time investment. Briefs will be published as part of Springer's eBook collection, with millions of users worldwide. In addition, Briefs will be available for individual print and electronic purchase. Briefs are characterized by fast, global electronic dissemination, standard publishing contracts, easy-to-use manuscript preparation and formatting guidelines, and expedited production schedules. Both solicited and unsolicited manuscripts are considered for publication in this series.

More information about this series at http://www.springer.com/series/8898

Árpád Furka

The Structure Dependent Energy of Organic Compounds

Árpád Furka
Eötvös Loránd University
Budapest, Hungary

ISSN 2191-5407 ISSN 2191-5415 (electronic)
SpringerBriefs in Molecular Science
ISBN 978-3-030-06003-9 ISBN 978-3-030-06004-6 (eBook)
https://doi.org/10.1007/978-3-030-06004-6

Library of Congress Control Number: 2018965448

This Springer imprint is published by the registered company Springer Nature Switzerland AG
The registered company address is: Gewerbestrasse 11, 6330 Cham, Switzerland

To Hossain Saneii
pioneer of automation of combinatorial
chemical syntheses

Preface

The subject of this book is the result of my endeavor to make organic chemistry better understood by my students. The reactions of organic compounds are influenced by two main factors. One factor is the energy of compounds the second one is their susceptibility to the attacks of reactants. The energy of organic compounds depends on their structure. It was a problem, however, that there were no data to show this dependency. The experimental heats of formation could not be applied for this purpose. This can be demonstrated by the heats of formation of some n-alkanes referring to gas state and expressed in kJ/mol.

CH_4	C_3H_8	C_5H_{12}	C_7H_{16}	C_9H_{20}
–74.9	–103.8	–146.4	–167.6	–229.0

It can be seen that there are considerable differences in the heats of formation despite the fact that these alkanes do not have specific structures that could be responsible for the differences. Only their composition differs and this is the reason of the differences in heats of formation.

The composition of isomeric compounds is the same, so the differences in their heats of formation can reflect the differences of their energy caused by differences in structure like in the case of cyclohexane and 1-hexene.

cyclohexane C_6H_{12}	1-hexene C_6H_{12}
$\Delta H^o_f = -123.1$	$\Delta H^o_f = -41.7$

Since the heats of formations in the great majority of cases do not make possible direct comparisons, the role of energy of reactants and products in the reactions is mainly neglected when teaching. Comparing the energies of unsaturated compounds by their heats of hydrogenations is an exception. As a consequence, when the reactivity of the organic compounds is considered almost exclusively their susceptibility to attacks by reagents is taking into account, and the energy factors are neglected.

In order to circumvent the problem caused by the traditional heats of formation in interpretation of the structure dependent energies of organic compounds an alternative thermochemical reference system was developed and published in 1983 [1]. The relative enthalpies replacing the heats of formation in the new system directly reflect the structure dependent energies of compounds and makes possible to compare them without any restriction. In the last three decades the relative enthalpies were successfully used in teaching organic chemistry for university students, and helped a deeper understanding of the properties of organic compounds.

Budapest, Hungary Árpád Furka

Reference

1. Furka Á (1983) Croatica Chemica Acta 56:199.

Contents

Abstract

The structure-dependent energies of organic compounds are parts of the conventional heats of formation. They are masked, however, by the large energies that are evolving as the consequence of the incorporation of the constituent elements into compounds. For this reason, the heats of formation heavily depend on the composition and, as a result, only the heats of formation of isomers can be directly compared. Heats of different reactions were used in comparisons to circumvent the problem. In the case of unsaturated hydrocarbons, for example, the heats of hydrogenations were compared. None of these approaches could be generally used, so demonstration of the structure-dependent energies remained complicated.

A general approach is offered to solve the problem by modifying the conventional thermochemical reference system. In this system, the reference substances, instead of the elements, are a series of compounds: unbranched alkanes and their properly selected derivatives (halides, ethers, sulfides, and tertiary amines). The zero energy assigned to the elements, of course, is no longer sustained. Nonzero energies are derived for them that refer to a single atom. In the new reference system, the energies (relative enthalpies) of the elements are (in kJ/mol): C −22.83, H_2 2 x 21.724, O_2 2 x 125.85, N_2 2 x (−89.3), S −41.626, F_2 2 x 198.0, Cl_2 2 x 47.11, Br_2 2 x 4.08, and I_2 −2 x 53.8.

The "relative enthalpies" of compounds that replace the heats of formation in the new system are calculated by adding to the heats of formation the relative enthalpies of the constituent elements multiplied by the number of their atoms in the molecule. The relative enthalpies express the structure-dependent relative energies of compounds and can be compared to each other without restrictions.

As examples, the relative enthalpies of selected representatives of the following types of compounds are presented: alkanes and cycloalkanes, alkenes and cycloalkenes, polyolefins and cyclic polyolefins, aromatic hydrocarbons, alcohols and phenols, ethers, peroxides, aldehydes and ketones, acetals, carboxylic acids, esters and anhydrides, alkyl, alkenyl and aryl halides, carboxylic acid halides, carbonyl halides, amines, carboxylic acid amides, hydrazine derivatives, nitriles, heteroaromatic compounds, nitro-compounds, organic nitrites and nitrates, organic sulfides, thiols, disulfides, sulfoxides, sulfones, organic sulfites, sulfates, selected inorganic compounds, organic radicals, organic cations and organic anions. Estimated stabilization energies of conjugated olefins, benzene, furan, pyrrole, and thiophene are also presented. The contribution of the transformed and newly formed compounds to the heats of reactions is also discussed.

Chapter 1
An Alternative Thermochemical Reference System

1.1 Introduction

The chemical compounds have energies that depend on their composition and structure. The absolute values of these energies are not known. The quantities that can be really measured are the heats of their reactions. The heat of reaction expresses the energy of the reaction products relative to that of the reactants. The traditional heats of formation coming out from these measurements are also relative values that compare the energy of compounds to the energies of arbitrarily selected reference substances. The reference substances are the elements that build up the compounds. By definition, the heats of formation of the elements are zero. Considering the elements as reference substances is a logical choice since the compounds are built up from elements. The heats of formation (ΔH_f^o) work perfectly for example in calculating the heat of any reaction (ΔH_r^o) if the heats of formation of both reactants and products are known. The heat of reaction can be calculated by subtracting the heats of formation of the reactants from the sum of the heats of formation of the products. Here is an example. Both reactants and the product are in gas phase [1, 2].

	$CH_3–CH{=}CH_2$	+ HCl =	$CH_3–CH_2Cl–CH_3$
ΔH_f^o	82.3	–23.5	–145.4

$$\Delta H_r^o = -145.4 - (82.3 - 23.5) = -204.2 \text{ kJ/mol}$$

It can be shown, however, that instead of zeros any other value can be assigned to the elements, and the result of calculation of the heat of reaction remains unchanged.

Electronic Supplementary Material The online version of this chapter (https://doi.org/10.1007/978-3-030-06004-6_1) contains supplementary material, which is available to authorized users.

Á. Furka, *The Structure Dependent Energy of Organic Compounds*, SpringerBriefs in Molecular Science, https://doi.org/10.1007/978-3-030-06004-6_1

Let assign 20 kJ/mol to H_2 (10 kJ per atom) 30 kJ/mol to carbon and 40 kJ/mol to Cl_2 (20 kJ per atom) and recalculate the heats of formation and the heat of reaction. It can be seen that despite of the assignment of arbitrary values to the elements, the value of the heat of reaction remains the same. The reason is very simple, the kinds and the number of atoms appearing in the reactants and in the products is the same, consequently the total of corrections appearing in both sides of the reaction equation is also the same and in calculation of the heat of reaction cancel out.

	$CH_3–CH=CH_2$ +	HCl =	$CH_3–CH_2Cl–CH_3$
	C_3H_6	HCl	C_3H_7Cl
Corrections C	90		90
Corrections H	60	10	70
Corrections Cl		20	20
Total correction	150	30	180
ΔH^o_f	82.3	–23.5	–145.4

Correcting the heats of formation by adding the corrections for C, H and Cl.

Corrected ΔH^o_f-s	232.3	6.5	34.6

The heat of reaction calculated from corrected heats of formation:

$$\Delta H^o_r = 34.6 - (232.3 + 6.5) = -204.2$$

The heat of formation of an organic compound is supposed to comprise two components:

1. A large part of the heat of formation of a compound arises from the transformation of its constituent atoms from the elemental state to the compound state. This part entirely depends on composition.
2. The second component comprises the energy coming from the structure of the compound. The first part of the heat of formation, however, is so large that completely masks the contribution arising from the structure. This can be demonstrated by comparing the heats of formation of two gas phase hydrocarbon compounds:

$$\text{2-methylpropane} \quad (C_4H_{10}, \Delta H^o_f = -134.5\,\text{kJ/mol})$$
$$\text{1-dodecene} \quad (C_{12}H_{24}, \Delta H^o_f = -165.4\,\text{kJ/mol})$$

By comparing the ΔH^o_f-s of the two compounds despite of its high energy double bond, 1-dodecene looks to have lower energy than the saturated 2-methylpropane.

It is useful to know—particularly when teaching chemistry—how the energy of organic compounds depends on structure. This can be demonstrated by comparing the structures and heats of formation of isomeric compounds. Since their composition is the same the heat arising from transformation of the elements to compound state is the same in the case of both compounds so they cancel out in comparisons. This makes possible comparing their heats of formation. The vast majority of compounds, however, are not isomeric and for this reason their heats of formation are totally non-comparable.

It was shown above, that different arbitrary values can be assigned to the heats of formation of elements without affecting the calculation of the heats of reactions. Keeping this in mind one may ask: could we assign appropriate values to the elements that eliminate the dependence of the heats of formation on composition? It will be shown that the answer is yes if the elements as reference substances are replaced by properly selected reference compounds.

1.2 Alkanes as Reference Substances

Structure-dependent energy comparisons are most meaningful if the data refer to gas state. The energy in gas state is free of the effects of intermolecular interactions. In order to find out whether the n-alkanes would be good choices for reference substances, the heats of a few hypothetical redistribution reactions are examined (Table 1.1) using the heats of formation listed in Table 1.3.

The data of Table 1.1 shows that the heats of redistribution reactions are practically zero. The only exception is the reaction where methane is formed. These results demonstrate that the gas phase n-alkanes from ethane up to at least eicosane are good candidates for reference substances. The energy of the reactants and the products is the same and does not depend on the number of carbon and hydrogen atoms present in the individual product molecules. This idea is further supported by the data in Table 1.2.

All the calculated heats of hydrogenolysis reactions, regardless of the starting n-alkane or the products, are practically the same, except again when methane is formed.

In Fig. 1.1 the enthalpies of formation of series of different organic compounds are plotted against their number of carbon atoms: n-alkanes and their derivatives with functional group at one of their chain end: alkenes, alkynes, aldehydes, alcohols and carboxylic acids.

If the first members of the series are left out of consideration, the results are descending parallel lines. Positions of the lines are determined by the functional

Table 1.1 Calculated heats of redistribution reactions of n-alkanes

Products	ΔH_r^o (kJ/mol)	Products	ΔH_r^o (kJ/mol)
$C_9H_{20} + C_{11}H_{24}$	0.1	$C_4H_{10} + C_{16}H_{34}$	0.0
$C_8H_{18} + C_{12}H_{26}$	0.1	$C_3H_8 + C_{17}H_{36}$	1.7
$C_7H_{16} + C_{13}H_{28}$	0.3	$C_2H_6 + C_{18}H_{38}$	0.1
$C_6H_{14} + C_{14}H_{30}$	0.1	$CH_4 + C_{19}H_{40}$	−10.6
$C_5H_{12} + C_{15}H_{32}$	0.2		

Reactants: $C_{10}H_{22} + C_{10}H_{22}$

Table 1.2 Calculated heats of hydrogenolysis reactions of some n-alkanes

Reactants	Products	ΔH_r^o (kJ/mol)
$C_{20}H_{42} + H_2$	$C_{10}H_{10} + C_{10}H_{10}$	43.9
$C_{20}H_{42} + H_2$	$C_{15}H_{32} + C_5H_{12}$	43.7
$C_{20}H_{42} + H_2$	$CH_4 + C_{19}H_{40}$	54.5
$C_{15}H_{32} + H_2$	$C_7H_{16} + C_8H_{18}$	43.2
$C_{15}H_{32} + H_2$	$CH_4 + C_{14}H_{30}$	54.3
$C_8H_{18} + H_2$	$C_4H_{10} + C_4H_{10}$	43.8

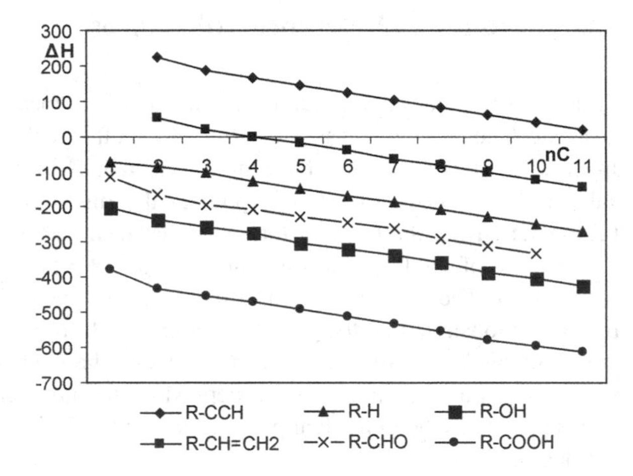

Fig. 1.1 Dependence of the heats of formation of series of gas state organic compounds on the number of carbon atoms (n_C)

groups. The high energy triple bond, for example, puts the alkynes line above those the other series and the lowest line belongs to the carboxylic acids.

Based on the above considerations the gaseous n-alkanes—except methane—are chosen as reference substances for hydrocarbons. The heats of formation are replaced by relative enthalpies denoted by H^o_{rel}, and zero relative enthalpy is assigned to all gaseous n-alkanes. The non-zero relative enthalpy of carbon (H^{C}_{rel}) and that of hydrogen (H^{H}_{rel}) is defined by Eq. 1.1. where n_{C} is the number of carbon atoms, H^{C}_{rel} and H^{H}_{rel} are the relative enthalpies of elemental carbon and hydrogen ($1/2H_2$), respectively.

$$0 = \Delta H_f^o + n_{\mathrm{C}} H^{\mathrm{C}}_{rel} + (2\mathrm{n_C} + 2) H^{\mathrm{H}}_{rel} \tag{1.1}$$

The linearity of the heat of formation of n-alkanes versus the number of carbon atoms curve, in almost the entire region, made possible to calculate the relative enthalpy of elemental carbon and hydrogen by linear regression from the heats of formation of the C5 to C20 n-alkanes using Eq. (1.2) (the rearranged form of Equation 1.1):

Table 1.3 Heats of formation and calculated relative enthalpies of gas state n-alkanes

Compound	Composition	ΔH_f^o	H_{rel}^o (kJ/mol)	Compound	Composition	ΔH_f^o	H_{rel}^o (kJ/mol)
Methane	CH_4	−74.9	−10.8	Undecane	$C_{11}H_{24}$	−270.3	−0.1
Ethane	C_2H_6	−84.7	0	Dodecane	$C_{12}H_{26}$	−290.9	0
Propane	C_3H_8	−103.8	1.5	Tridecane	$C_{13}H_{28}$	−311.5	0
Butane	C_4H_{10}	−126.1	−0.2	Tetradecane	$C_{14}H_{30}$	−332.1	0
Pentane	C_5H_{12}	−146.4	0.1	Pentadecane	$C_{15}H_{32}$	−352.8	−0.1
Hexane	C_6H_{14}	−167.2	0	Hexadecane	$C_{16}H_{34}$	−373.3	0
Heptane	C_7H_{16}	−187.6	0.2	Heptadecane	$C_{17}H_{36}$	−393.9	0.1
Octane	C_8H_{18}	−208.4	0	Octadecane	$C_{18}H_{38}$	−414.6	0
Nonane	C_9H_{20}	−229.0	0	Nonadecane	$C_{19}H_{40}$	−435.1	0.1
Decane	$C_{10}H_{22}$	−249.7	−0.1	Eicosane	$C_{20}H_{42}$	−455.5	0.3

$$-\Delta H_f^o = n_{\mathrm{C}}(H_{rel}^{\mathrm{C}} + 2H_{rel}^{\mathrm{H}}) + 2H_{rel}^{\mathrm{H}} \tag{1.2}$$

The calculated values are, $H_{rel}^{\mathrm{C}} = -22.83\,\mathrm{kJ/mol}$ and $H_{rel}^{\mathrm{H}} = 21.724$. Since H_{rel}^{H} refers to ½H_2, the H_{rel}^{H} of H_2 is 43.448.

The relative enthalpy of any hydrocarbon (H_{rel}^o) can be calculated by correcting its heat of formation using the non-zero values of H_{rel}^{C} and H_{rel}^{H}:

$$H_{rel}^o = \Delta H_f^o - 22.83n_{\mathrm{C}} + 21.724n_{\mathrm{H}} \tag{1.3}$$

The relative enthalpies of twenty gas state n-alkanes calculated from their heats of formation are summarized in Table 1.3. The values are zero or very close to it as expected for the reference compounds.

Figure 1.2 shows that the line of the relative enthalpy of n-alkanes is at zero and the corrected heats of formation of compounds within all series—except the first members of the series—are the same. The corrected heats of formation no longer depend on the number of carbon atoms.

1.3 Compounds Containing O, N, S, F, Cl, Br and/or I

In order to extend the reference system to compounds containing—in addition to carbon and hydrogen—one or more atoms of other elements most frequently occurring in organic compounds, further groups of reference compounds were selected:

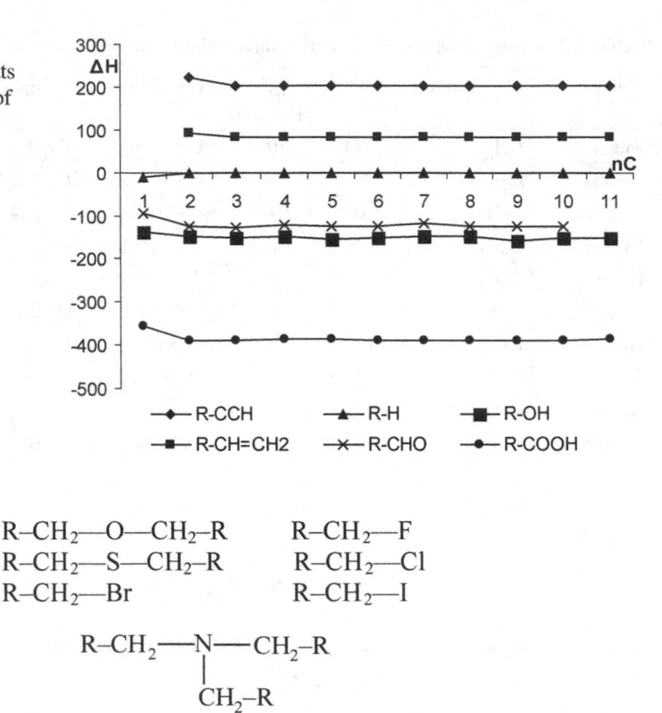

Fig. 1.2 Plotting the corrected values of the heats of formation of the series of compounds against the number of carbon atoms

$R–CH_2—O—CH_2–R$ $R–CH_2—F$
$R–CH_2—S—CH_2–R$ $R–CH_2—Cl$
$R–CH_2—Br$ $R–CH_2—I$

$R–CH_2—N(CH_2–R)—CH_2–R$

R means n-alkyl group. Methylene group is included into the structures to eliminate the first members of the series of compounds.

The relative enthalpies of these compounds—already corrected for their C and H atoms—were adjusted to zero by assigning proper values to O, N, S, F, Cl, Br and I. This is exemplified with the gas phase dialkyl ethers. The available heats of formation of the members (but excluding the first member dimethyl ether) were first corrected with H^o_{rel}-s of C and H according to Eq. (1.3). The mean value of the corrected heats of formation was −125.85 kJ/mol. The H^o_{rel} values of the reference ethers could be made zero, or values very close to zero, by adding 125.85 kJ/mol to the heats of formation of these compounds, already corrected by H^o_{rel}-s of C and H. This means that 125.85 kJ/mol was assigned to the H^o_{rel} of oxygen.

Similar calculations were carried out with the reference compounds containing N, S, F, Cl, Br and I. The values of so determined H^o_{rel}-s of the elements, including those of the previously known H^o_{rel}-s of carbon and hydrogen are shown below.

C: −22.83 **H**: 21.724 **O**: 125.85 **N**:−89.3 **S**:−41.626

F: 198.0 **Cl**: 47.11 **Br**: 4.08 **I**: −53.8

The H^o_{rel} of diamond is different: −20.94 kJ/mol

The application of these values for calculation of the relative enthalpy of a compound is demonstrated using benzoyl chloride as example.

Composition of gas state benzoyl chloride and its heat of formation are: C_7H_5ClO and −109.2 kJ/mol, respectively.

Corrections:	C	7 x −22.83 = −159.81
	H	5 x 21.724 = 108.62
	O	125.85
	Cl	47.11
Total of corrections:		121.77

The total of corrections is added to the heat of formation to give the H^o_{rel} of benzoyl chloride:

$$H^o_{rel} = -109.2 + 121.77 = 12.57\,\text{kJ/mol}$$

Finally it is demonstrated by an example, that replacement of the heats of formation by relative enthalpies does not change the result of calculation of the heat of reaction.

$$CH_3–CH=CH_2 + HCl = CH_3–CHCl–CH_3$$

ΔH^o_f	20.4	−92.3	−146.4
ΔH^o_r (from ΔH^o_f −s) = −74.5 kJ/mol			
H^o_{rel}	82.3	−23.5	−15.7
ΔH^o_r (from H^o_{rel} −s) = −74.5 kJ/mol			

As mentioned above, the heats of formation are not suitable to compare the structure dependent energies of organic compounds because they include the energy that evolves when the elements enter from their elemental state to the compound state. This energy is eliminated from the relative enthalpies of compounds by adding the relative enthalpies of the component elements to the heats of formation. This can be demonstrated using the hydrogenolysis reactions listed in Table 1.2. Hydrogen is added to alkanes and two alkanes are formed in each case. Since the two alkanes form from the reactant alkane by inclusion of the two hydrogen atoms into the two products, the heat of the reaction can be considered as the energy that evolves when the hydrogen enters into compound (alkane) state. The heats of the reactions when no methane is formed are between −43.2 and −43.9 kJ/mol. These values refer to two hydrogen atoms so the values for one hydrogen atom is −21.6 to −21.9 kJ/mol (mean value, −21.75 kJ/mol). This −21.75 kJ/mol is the heat of the reaction when one elemental hydrogen atom enters into alkane state so the relative enthalpy assigned to hydrogen, 21.724 kJ/mol seems to be the right value to compensate for it.

The alternative thermochemical reference system outlined above is by no means intended to replace the traditional one. The primary importance of the new system is that it makes possible to mine out from the heats of formation the energies of organic compounds that depend exclusively on their structure.

Most of the heats of formation of organic and inorganic compounds that were used to calculate the H^o_{rel} appearing in this book were gathered from the compilations of Cox and Pilcher [3], Stull et al. [4] and NIST [5]. Entropies of compounds were taken from Stull et al. [4]. The heats of formation of organic radicals, cations and anions came from data collections of Argonne National Laboratory [6] and Goos et al. [7]. If more than one heat of formation was available for a given compound the newest value was used for calculation the H^o_{rel}. The original heats of formation of a large range of compounds as well as their calculated relative enthalpies are found in the Supplement.

References

1. Furka Á (1983) Croat Chem Acta 56:199
2. Furka Á, Czimer I (1993) Struct Chem 4:327
3. Cox JD, Pilcher G (1970) Thermochemistry of organic compounds. Academic Press, London, New York
4. Stull DR, Westrum EF Jr, Sinke GC (1969) The chemical thermodynamics of organic compounds. Wiley, New York
5. http://webbook.nist.gov/chemistry
6. Argonne National Laboratory, Active thermochemical tables, version 1.112
7. Goos E, Burcat A, Ruscic B (2010) Extended third millennium ideal gas and condensed phase thermochemical database for combustion with updates from active thermochemical tables

Chapter 2
Hydrocarbons

2.1 Alkanes

The relative enthalpies of the gas state n-alkanes are listed in Table 1.3. Their values except that of methane are practically zero. It is long known from comparisons of the heats of formation of isomeric alkanes that branching mostly causes stabilization. This is supported by the relative enthalpies of Table 2.1. Substitutions by methyl group at the chain end bring about 7–8 kJ/mol energy reduction. Double substitution on the same carbon atoms reduces the energy to −17, −18 kJ/mol [1].

It seems reasonable to call attention to a fact. The H^{o}_{rel} of 2,2-dimethylpropane is −19.5 kJ/mol. This value is very close to the H^{o}_{rel} of diamond −20.94 kJ/mol. It is possible that this is not an accidental similarity since the four bonds of the central carbon atom of 2,2-dimethylpropane are formed with four other carbon atoms. This is similar to the four bonds made by all carbon atoms of diamond (except those at the surface) and the reduced energy of 2,2-dimethylpropane may come from the contribution of the diamond-like carbon atom.

Table 2.1 also shows that if the methyl substitution is on the third or fourth carbon atom from the chain end instead of the second one, the energy reduction is lower, around 3.7–4.5 kJ/mol. In the case of crowded substitutions inside the chains, the energy reducing effect of branching is strongly decreased. The 2,2,5,5-tetramethyl hexane is an exception since in its H^{o}_{rel} the effect of all the four methyl groups shows up, certainly because the double substitutions occur on the second carbon atom from both ends of the chain.

Dependence of the energy of alkanes on their structure is known for a long time. Comparing the heats of formation of isomeric alkanes leads to the same results reflected by their H^{o}_{rel}-s. This is exemplified below.

Á. Furka, *The Structure Dependent Energy of Organic Compounds*, SpringerBriefs in Molecular Science, https://doi.org/10.1007/978-3-030-06004-6_2

Table 2.1 Effect of branching on the energy of alkanes

Compound	H^o_{rel} (kJ/mol)	Compound	H^o_{rel} (kJ/mol)
2-Methylpropane	−8.6	2,2-Dimethylbutane	−18.4
2-Methylbutane	−8.0	2,2-Dimethylpentane	−18.3
2-Methylpentane	−7.1	2,2-dimethylhexane	−16.3
2-Methylhexane	−7.1	2,2-Dimethylheptane	−17.9
2-Methylheptane	−7.1	2,2-Dimethyloctane	−17.9
2-Methyloctane	−7.2	3-Ethyl-3-Methylpentane	−6.6
3-Methylpentane	−4.4	4-Ethyl-3,3-dimethylhexane	−4.1
3-Methylhexane	−4.5	3-Ethyl-2,2,3-trimethylpentane	−3.0
3-Methylheptane	−4.2	4-Ethyl-2,3-dimethylhexane	−3.2
3-Methyloctane	−4.3	3-Ethyl-3,4-dimethylhexane	−1.7
3-Methylnonane	−4.4	2,2,3,3-Tetramethylhexane	−8.3
4-Methyl heptane	−3.7	2,3,4,5-Tetramethylhexane	−8.4
2,2-Dimethylpropane	−19.5	2,2,5,5-Tetramethylhexane	−35.7

$CH_3–CH_2–CH_2–CH_2–CH_3$ $CH_3–CH(CH_3)–CH_2–CH_3$ $CH_3–C(CH_3)_2–CH_3$

ΔH^o_f	−146.4	−154.5	−166.0
ΔH^o_f difference:	−8.1	11.5	
H^o_{rel}	−0.1	−8.0	−19.5

It can be seen that in the case of isomeric compounds, comparison of ΔH^o_f-s and H^o_{rel}-s leads to the same result (small differences may occur because of rounding in calculations). What makes the difference is, that similar results can be achieved by comparing the H^o_{rel}-s of non-isomeric compounds while their ΔH^o_f-s are non-comparable.

$CH_3–CH_3$ $CH_3–CH(CH_3)–CH_3$ $CH_3–C(CH_3)_2–CH_3$

ΔH^o_f	−84.7	−134.5	−166.0
ΔH^o_f difference:	−49.8	−81.3	
H^o_{rel} =	0.0	−8.6	−19.5

2.2 Cycloalkanes

Table 2.2 shows that the energy of cycloalkanes depends on the ring member. As can be seen, the H^o_{rel} of cyclohexane is close to zero like those of the n-alkanes. The H^o_{rel} of cyclobutane and cyclopropane is very high and this reflects the high strain in these rings. The H^o_{rel} per the number of carbon atoms (H^o_{rel}/n_C) better reflects the ring strain. The highest value belongs to cyclopropane and somewhat lower to cyclobutane. The H^o_{rel}/n_C-s of cyclopentane, cycloheptane and cyclooctane are relatively low values far from those of cyclopropane and cyclobutane. Nevertheless they may also reflect some strains in these molecules.

The alkyl substitution on ring carbons brings about stabilization like in the case of open chain alkanes. Table 2.3 shows stabilizations brought about by methyl substitutions on ring carbons. Since the ΔH^o_f, and as a consequence the H^o_{rel}, of the gas phase methylcyclopropane is not available to compare, the highest stabilization is observed at the strained methylcyclobutane.

Considering the substituent effects in Table 2.3 and in all the subsequent ones the positive values mean energy increasing effects while the negative ones reflect stabilization.

Substitution of cyclopentane shows that stabilization is not significantly affected by the length of the substituent. This phenomenon is even more supported by the data of Table 2.4. Double substitution on the same carbon atom is effective like at the open chain hydrocarbons. Steric effect is also reflected by the H^o_{rel} of cis-1,2-dimethylcyclohexane.

Table 2.2 Relative enthalpy of cycloalkanes

Compound	H^o_{rel} (kJ/mol)	H^o_{rel}/n_C (kJ/mol)	Compound	H^o_{rel} (kJ/mol)	H^o_{rel}/n_C (kJ/mol)
	115.2	38.3		0.6	0.1
	109.2	27.3		25.0	3.6
	25.9	5.2	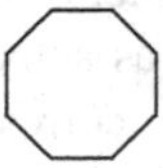	39.1	4.9

Table 2.3 Relative enthalpies of substituted cycloalkanes

Compound	H^o_{rel} (kJ/mol)	Effect (kJ/mol)	Compound	H^o_{rel} (kJ/mol)	Effect (kJ/mol)
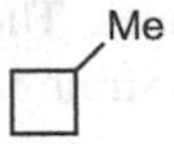 Me	82.2	−27	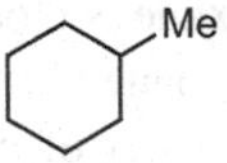 Me	−10.5	−11.1
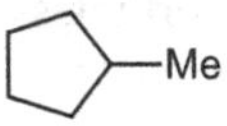 Me	17.0	−8.9	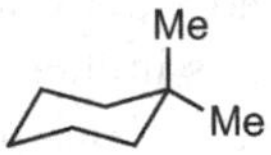 Me Me	−16.0	−16.6
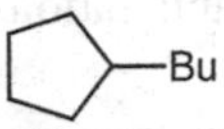 Bu	17.3	−8.6	 Me Me	−15.0	−15.6
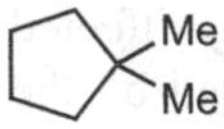 Me Me	6.1	−19.8	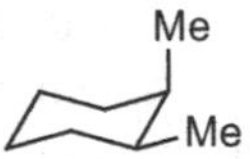 Me Me	−7.2	−7.8

Table 2.4 Relative enthalpies of monosubstituted cyclopentanes and cyclohexanes

Compound	H^o_{rel} (kJ/mol)	Compound	H^o_{rel} (kJ/mol)
Ethylcyclopentane	17.3	Ethylcyclohexane	−6.8
Propylcyclopentane	16.9	Propylcyclohexane	−7.7
Pentylcyclopentane	17.3	Pentylcyclohexane	−7.0
Hexylcyclopentane	17.3	Hexylcyclohexane	−6.9
Heptylcyclopentane	17.3	Heptylcyclohexane	−6.9
Octylcyclopentane	17.4	Octylcyclohexane	−6.9
Nonylcyclopentane	17.4	Nonylcyclohexane	−6.9
Decylcyclopentane	21.5	Decylcyclohexane	−6.9

Table 2.5 shows the relative enthalpies of polycycloalkanes. The H^o_{rel}-s of the two decalin isomers reflect the effect of their two tertiary carbon atoms. The fusion carbon atoms can be considered as substituted carbon atoms, and so the H^o_{rel}-s of the decaline isomers roughly correspond to the H^o_{rel}-s of the cis- and trans-1,2-dimethylcyclohexane (Table 2.3).

Table 2.5 Relative enthalpies of polycycloalkanes

Compound	H^o_{rel} (kJ/mol)	Compound	H^o_{rel} (kJ/mol)	H^o_{rel}/n_C
	−19.4	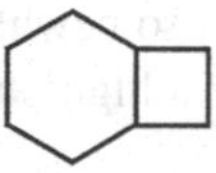	94.8	
	−6.5		256.1	64.0
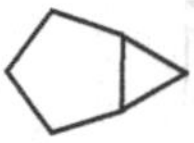	118.3		244.8	49.0
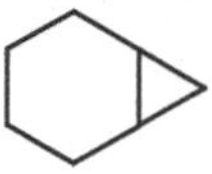	102.3	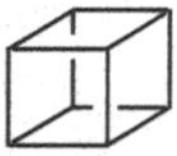	588.2	73.5

The other bicycloalkanes in the table have one strained ring and, as a consequence, their relative enthalpy is considerably increased. The cyclopropane ring fused to cyclopentane and cyclohexane brings about 92.4 and 101.7 kJ/mol energy increase, respectively, that is close to the H^o_{rel} of cyclopropane (115.2 kJ/mol) particularly if the potential energy reducing effects of the two tertiary fusion carbon atoms are taking into account. The situation is similar if the cyclobutane ring is fused to cyclohexane.

The H^o_{rel} of bicyclo[4,2,0]octane can be compared to those of the decaline isomers (in order to take into account the effect of the fused carbons). The so calculated differences (114.2 and 101.3 kJ/mol) are again close to the H^o_{rel} of cyclobutane (109.2 kJ/mol). The H^o_{rel}-s of bicyclobutane, spiropentane and cubane are very high. These values reflect extremely high strains in these molecules. Their H^o_{rel}/n_C values show that the strain increases from spiropentane through bicyclobutane to cubane.

The H^o_{rel} of bicyclobutane is higher than the sum of that of two cyclopropanes. This may reflect the high strain on the two fusion carbon atoms.

The H^o_{rel} of spiropentane is also somewhat higher than the sum of the H^o_{rel}-s of two cyclopropanes that may show a high strain on the central carbon atom.

The H^o_{rel} of cubane looks extremely high particularly if compared to the sum of the H^o_{rel}-s of two cyclobutanes. This may reflect that the bonding structure of all the eight carbon atoms is very unfavorable.

2.3 Alkenes

2.3.1 *Mono-olefins*

The increased energy of ethylene and of other alkenes is known for a long time. The magnitude of the increase was not clear, however, because there was no proper other compound with the same composition to compare with. For this reason the heat of hydrogenation seemed to give a measure for the increased energy. The heat of hydrogenation that proved very useful for comparing the energies of different alkenes, did not give an adequate value for the increased energy of ethylene. The reason is that the heat of hydrogenation is composed from two components: the structure dependent increased energy and the energy evolving when the elemental hydrogen enters the compound state. This later value can be taken from Table 1.2. The mean value of the heat of hydrogenolysis reactions when no methane is formed is −43.65 kJ/mol. The heat of the hydrogenation of ethylene calculated from the ΔH^o_f-s of the reactants and the product is −137.0 kJ/mol.

$$CH_2{=}CH_2 + H_2 = CH_3{-}CH_3$$

The part of the heat of reaction responsible for transforming the double C–C bond into single C–C bond can be calculated by subtracting from the heat of hydrogenation reaction (−137 kJ/mol) the heat of the hydrogenolysis reaction: −43.65 kJ/mol. The result is −93.35 kJ/mol. So the structure dependent energy of ethylene calculated this way is this value with positive sign: 93.35 kJ/mol. The H^o_{rel} of ethylene taken from Table 2.6, is 93.5 kJ/mol. The two values are practically the same, demonstrating that the H^o_{rel} of ethylene really reflects the structure dependent energy.

Table 2.6 Relative enthalpies of 1-alkenes

Compound	H^o_{rel} (kJ/mol)	Stabilization (kJ/mol)	Compound	H^o_{rel} (kJ/mol)	Stabilization (kJ/mol)
Ethylene	93.5		1-Dodecene	82.2	−11.3
Propene	82.3	−11.2	1-Tridecene	82.0	−11.5
1-Butene	82.4	−11.1	1-Tetradecene	82.2	−11.3
1-Pentene	82.2	−11.3	1-Pentadecene	82.1	−11.4
1-Hexene	82.0	−11.5	1-Hexadecene	82.1	−11.4
1-Heptene	82.0	−11.5	1-Heptadecene	82.1	−11.4
1-Octene	82.0	−11.5	1-Octadecene	82.1	−11.4
1-Nonene	82.1	−11.4	1-Nonadecene	82.2	−11.3
1-Decene	82.1	−11.4	1-Eicosen	82.2	−11.3
1-Undecene	82.0	−11.5			

The data of Table 2.6 also demonstrate that alkyl substituents have energy reducing, that is stabilization effects of around 11.4 kJ/mol that are independent of the chain length. This effect is stronger than that in the open chain saturated hydrocarbons (~8 kJ/mol).

The effect of multi substitution is demonstrated in Table 2.7. Stabilization by double substitution on the same carbon atom of ethylene is 28.9 kJ/mol.

Table 2.7 Effect of substituting the hydrogens of ethylene by methyl groups

Compound	H^o_{rel} (kJ/mol)	Stabilization (kJ/mol)	Compound	H^o_{rel} (kJ/mol)	Stabilization (kJ/mol)
$H_2C{=}$ CH_3 CH_3	64.6	−28.9	H_3C CH_3 CH_3	61.6	−31.9
H_3C CH_3	75.5	−18.0	H_3C CH_3 H_3C CH_3	53.4	−40.1
H_3C CH_3	71.3	−22.2			

Contribution per methyl group is 14.5 kJ/mol, this is higher than stabilization by a single substituent. Triple or tetra substitution brings about a total effect of 31.9 and 40.1 kJ/mol (10.6 and 10 kJ/mol/number of methyl groups, respectively).

If mono-methyl substitution occurs on both carbon atoms, the magnitude of stabilization depends on the geometry of the molecule.

2.3.2 *Cyclic Mono-alkenes*

The effect of introducing a double bond into the rings is shown by comparing the H^o_{rel} values of Table 2.8 to those of cycloalkanes of Table 2.2. The increments relative to the corresponding cycloalkanes are also found in the Table. The double bonds in cyclic alkenes are part of a cyclic chain so their H^o_{rel}-s should be compared to the H^o_{rel} of cis-2-butene (75.5 kJ/mol).

Table 2.8 Relative enthalpy of cyclic monoolefins

Compound	H^o_{rel} (kJ/mol)	Effect (kJ/mol)	Compound	H^o_{rel} (kJ/mol)	Effect (kJ/mol)
	293.5	178.3	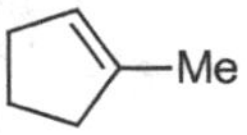 –Me	74.9	−17.7
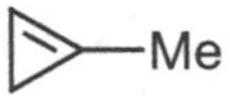 –Me	283.0	−10.5		74.9	74.3
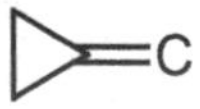 =C	240.0		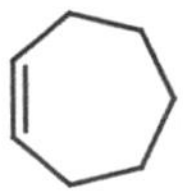	91.6	66.6
	196	86.8		93.2	54.0
	95.6	69.7			

Table 2.9 Effect of double bond conjugation

Compound	H^o_{rel} (kJ/mol)	For comparison	H^o_{rel} (kJ/mol)	Stabilization (kJ/mol)
C=C	93.5			
C=C–C=C	149.2	C=C + C=C	187.0	−37.8
		C=C–C + C=C–C	164.6	−15.4
		C=C–C–C=C	165.0	−15.8
=C C=C C=C	202.1	C=C–C + C=C–C + C–C =C–C trans	235.9	−33.8
C=C C=C C=C	206.7	C=C–C + C=C–C + C–C=C–C cis	240.1	−33.4

The increment in the case of cyclopropene is very high, nearly 180 kJ/mol. This value is more than twice as high as the H^o_{rel} of 2-butene. This is certainly the consequence of the extremely high strain in the molecule.

The cyclohexene increment is about the same as the H^o_{rel} of 2-butene as expected. The increments of the other cyclic mono-alkenes are unexpectedly lower than that. Cyclobutene, however, is an exeption.

The H^o_{rel} of methylenecyclopropane is considerable higher than the sum of the H^o_{rel}-s of ethylene and cyclopropane (208.7 kJ/mol) that seems to indicate the particularly strained state of the central carbon atom.

Methyl groups on the double bonded carbon have stabilization effect close to those observed in open chain alkenes (10–18 kJ/mol).

2.3.3 *Di and Polyolefins. Stabilization Energy of Butadiene*

It is well known that double bonds in polyalkenes that occupy conjugated position reduce the energy. This was supported by the experimental finding that the heat of hydrogenation of butadiene was 38 kJ/mol less, relative to that of two moles of ethylene. This difference could be considered as stabilization energy of butadiene.

The relative enthalpies provide new possibilities to estimate the stabilization energy. The first possibility is to compare the H^o_{rel} of butadiene to the sum of the H^o_{rel}-s of two ethylene molecules (Table 2.9). In this case stabilization energy would be 37.8 kJ/mol like in comparing the heats of hydrogenations. This comparison, however, can't be justified since both double bonded carbon atoms in butadiene are connected by a C–C bond which in itself has a stabilization effect. These C–C bonds are missing from the two ethylene molecules. It is better to compare the H^o_{rel} of butadiene to the sum of the H^o_{rel}-s of two propene molecules since each propene is stabilized by a C–C bond. The energy difference of the sum of the H^o_{rel}-s of two propene molecules and the H^o_{rel} of butadiene is 15.4 kJ/mol.

Another possibility for estimation of stabilization energy is to compare the H^o_{rel} of butadiene with that of 1,4-pentadiene. In this compound both double bonded parts of the molecule are stabilized by a C–C bond. The result is 15.8 kJ/mol. The two values, 15.4 and 15.8 obtained by two different approaches are very close. So the stabilization energy of butadiene, that comes from the conjugated position of two double bonds is estimated to be 15–16 kJ/mol. Based on similar considerations stabilization energies of the trans- and cis-1,3,5-hexatrienes are 33.8 and 33.4 kJ/mol, respectively.

In Table 2.10 the relative enthalpies of cycloalkenes containing conjugated double bonds are summarized. The H^o_{rel}-s of the four dienes in the first column are close to each other and close to that of butadiene (149.2 kJ/mol). Each of these molecules contains two additional stabilizing C–C bonds. The H^o_{rel} of cis-1,3-pentadiene (137.8 kJ/mol in Table 2.10) if compared to that of butadiene (149.2 kJ/mol) shows that one such C–C bond reduces the energy by 11.4 kJ/mol. Thus the H^o_{rel}-s of the four

Table 2.10 Conjugated double bonds in cycloalkene rings

Compound	H^o_{rel} (kJ/mol)	Compound	H^o_{rel} (kJ/mol)
	150.1		195.9
	145.2	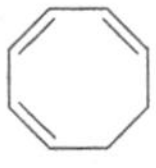	228.4
	150.7		289.2
	157.1	C=C–C=C–C	137.8

cycloalkadienes taking into consideration the effect of the two additional stabilizing C–C bonds are expected to be around 126–127 kJ/mol. The H^o_{rel}-s are considerably higher, indicating that the effect of the two C–C bonds are probably overcompensated by ring strains. The two double bonds are the best and the worst accommodated in the six and eight member rings, respectively.

The H^o_{rel} of cycloheptatriene is somewhat lower than those of the open chain hexatrienes (~202, ~206 kJ/mol Table 2.12). They are not as low as expected (around 180 kJ/mol) considering the two C–C bonds connecting the two ends of the triene system in the ring. The H^o_{rel} of cyclooctatriene is even higher. This shows that the accommodation of the three conjugated double bonds in the eight member ring is even more unfavorable than that in the seven member one.

Cyclooctatetraene has four conjugated double bonds. It is a question whether or not there is a sign of a special stabilization relative to the open chain counterparts. The molecule has eight –CH= groups and it is supposed that the H^o_{rel} of the molecule is the sum of the contribution of the –CH= groups. The H^o_{rel} values for one –CH= group can be estimated by comparing the H^o_{rel} of butadiene to that of ethylene.

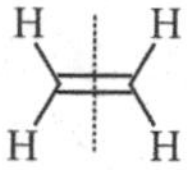

93.5 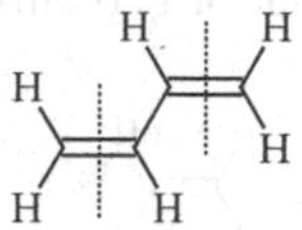 149.2

What remains after subtracting the H^o_{rel} of ethylene from that of butadiene can be considered as the contribution of the two –CH= groups: 149.2 − 93.5 = 55.7 kJ/mol (27.85 kJ/mol/group). Using the 27.85 kJ/mol value for estimating the H^o_{rel} of cyclooctatetraene 222.8 kJ/mol comes out. The H^o_{rel} of cyclooctatetraene in Table 2.10 is much higher than that showing that there is no special stabilization. The high H^o_{rel} (289.2 kJ/mol) rather reflects the unfavorable accommodation of the four conjugated double bonds in the eight member ring.

Since the H^o_{rel} of cyclooctatetraene does not demonstrate a special stabilization it is a question whether or not the stabilizing effects of the C–C single bonds show up. In this case the H^o_{rel} of cyclooctatetraene needs to be compared to the sum of the H^o_{rel}-s of four 2-cyclobutenes.

H_3C CH_3 4×75.5 = 302 H_3C CH_3 71.3x4 = 285.2 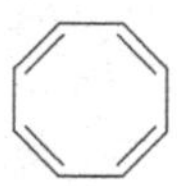 289.2

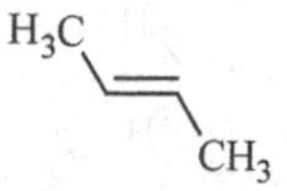

The H^o_{rel} of cyclooctatetraene nearly corresponds to the H^o_{rel}-s of four trans-2-butenes reflecting the stabilizing effects of the C–C bonds.

Table 2.11 Relative enthalpies of allenes

Compound	H^o_{rel} (kJ/mol)	Stabilization (kJ/mol)	Compound	H^o_{rel} (kJ/mol)	Stabilization (kJ/mol)
C=C=C	210.5		C–C=C=C–C	198.1	−12.4
C=C=C–C	201.2	−9.3	C=C=C(C)–C	188.1	−22.4
C=C=C–C–C	205.2	−5.3			

2.4 Allenes

Allene is known a high energy compound as well as its substituted derivatives. This is reflected by the H^o_{rel}-s of Table 2.11. The H^o_{rel} of allene considerably exceeds not only that of one but even that of two ethylene molecules.

$H_2C{=}CH_2$ $2\times93.5 = 187$

$H_2C{=}C{=}CH_2$ 210.5

This indicates the particularly strained state of the central carbon atom. The rest of the H^o_{rel}-s in the table show the stabilizing effects of alkyl groups.

2.5 Alkynes

In addition to alkenes and allenes, acetylene and the other alkynes are also known as high energy compounds. The H^o_{rel} of acetylene exceeds even that of allene demonstrating the very unfavorable effect of the triple bond,

$H_2C{=}C{=}CH_2$ 210.5

H–C≡C–H   224.5

As Table 2.12 shows the stabilization effect of alkyl groups is particularly strong, around 20 kJ/mol/group. As it can be seen the H^o_{rel} of 1,3-butadiyne is less by 24.1 kJ/mol than that of two acetylenes. Because of the extra C–C bond in 1,3-butadiyne, it seems more appropriate to compare the H^o_{rel} of this molecule to that of two propynes (407.6 kJ). In contrast to butadiene, 1,3-butadiyne does not

Table 2.12 Relative enthalpy of alkynes

Compound	H^o_{rel} (kJ/mol)	Stabilization (kJ/mol)	Compound	H^o_{rel} (kJ/mol)	Stabilization (kJ/mol)
C≡C	224.5		C–C≡C–C	185.3	−39.2
C≡C–C	203.8	−20.7	C–C≡C–C–C	188.5	−36.0
C≡C–C–C	204.2	−20.3	C≡C–C≡C	424.9	−24.1
C≡C–C–C–C	203.9	−20.3			
2x C≡C–C	407.6		C≡C–C≡C	424.9	

show extra stabilization. Since its H^o_{rel} is higher by about 17 kJ/mol than that of two molecules of propynes rather destabilization is detected.

2.6 Aromatic Hydrocarbons

2.6.1 Stabilization Energy of Benzene

The simplest aromatic hydrocarbon is benzene. It is well known to show extra stabilization when compared to conjugated polyolefins. Stabilization energy was defined as the difference of the heat of formation of benzene and that of a hypothetic reference compound, 1,3,5-cyclohexatriene. This compound, however, does not exist. So the stabilization energy could not be determined experimentally. In order to circumvent this difficulty the ΔH^o_f of the reference compound was calculated from empirically or theoretically derived bond energies. Following this approach Klages [2] and Dewar [3] suggested 150.2 and 83.7 kJ/mol for stabilization energy, respectively. Furka and Sebestyén [4] suggested three closely similar stabilization energies: 82.3, 90.7 and 96.2 kJ/mol calculated from the ΔH^o_f-s of benzene, ethylene, 1,3-butadiene and 1,3,5-hexatriene.

The relative enthalpy of benzene is 76.3 kJ/mol. This is its structure dependent energy that can be compared to the structure dependent energy of any other organic compound. It is unnecessary to use in comparisons data of hypothetical compounds. Taking into account that benzene is three times unsaturated, it is indeed meaningful to compare the following two H^o_{rel}-s:

$$\begin{array}{ll} \text{benzene} & 76.3\,\text{kJ/mol} \\ \text{ethylene} & 93.5\,\text{kJ/mol} \end{array}$$

The structure dependent energy of benzene is lower than that of a single ethylene molecule containing a single unsaturated C=C bond. No need to compare it to three ethylene molecules or to hexatrienes.

Table 2.13 Relative energy of benzene and of its n-alkyl substituted derivatives

Compound	H^o_{rel} (kJ/mol)	Effect (kJ/mol)	Compound	H^o_{rel} (kJ/mol)	Effect (kJ/mol)
Benzene	76.3		Nonylbenzene	62.0	−14.3
Toluene	64.0	−12.3	Decylbenzene	62.0	−14.3
Ethylbenzene	64.4	−11.9	Undecylbenzene	62.1	−14.2
Propylbenzene	63.0	−13.3	Dodecylbenzene	62.1	−14.2
Butylbenzene	62.0	−14.3	Tridecylbenzene	62.2	−14.2
Pentylbenzene	62.1	−14.2	Tetradecylbenzene	62.1	−14.2
Hexylbenzene	62.1	−14.2	Pentadecylbenzene	62.1	−14.2
Heptylbenzene	62.1	−14.2	Hexadecylbenzene	62.2	−14.1
Octylbenzene	62.1	14.2			

If one sticks to the concept of stabilization energy, the method already used in the case of cyclooctatetraene can be applied taking into account that benzene contains six –CH= groups.

one −CH= group of butadiene:	27.85 kJ/mol
six −CH= groups 6 × 27.85	167.1 kJ/mol
H^o_{rel} of benzene	76.3 kJ/mol
Difference	90.8 kJ/mol

The difference (90.8 kJ/mol) can be considered as stabilization energy of benzene. The 167.1 kJ/mol H^o_{rel} would belong to a hypothetical carbon ring comprising three conjugated double bonds. Comparing to this the H^o_{rel} of benzene, 76.3 kJ/mol, is very low showing the very high energy reducing effect of aromaticity.

2.6.2 Substituted Benzene Derivatives

The H^o_{rel} of benzene is considerably reduced by substituting its hydrogen atoms by alkyl groups. Table 2.13 demonstrates that the stabilizing effect of an alkyl group is around 12–14 kJ/mol and this effect is not influenced by the length of the group.

If the hydrogen atoms of benzene are gradually substituted by alkyl groups the energy drops in each stage. This is demonstrated by the data of Table 2.14, where the substituents are methyl groups. The H^o_{rel} of the derivatives drops step by step. The H^o_{rel} of hexamethylbenzene is 11.4 kJ/mol close to zero.

It is worthwhile to consider: the hexamethylbenzene formally contains three double bonds and, despite this, its energy is close to those of n-alkanes. What is responsible of this is aromaticity with about energy reducing effect of about 90 kJ/mol, and the effect of substitutions that contributes to reducing of the energy by an additional 65 kJ/mol. The data of Table 2.15 shows that the H^o_{rel} of alkyl derivatives depends on the structures of the alkyl groups, too.

Table 2.14 Relative energy alkyl substituted derivatives of benzene

Compound	H^o_{rel} (kJ/mol)	Compound	H^o_{rel} (kJ/mol)
	76.3		
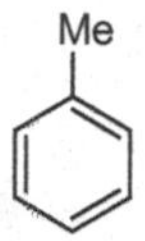	64.0	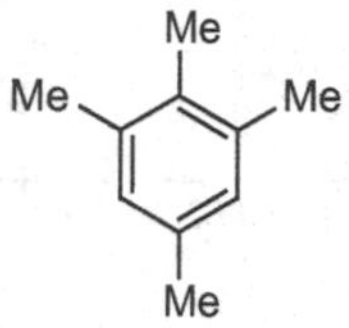	31.0
	52.6	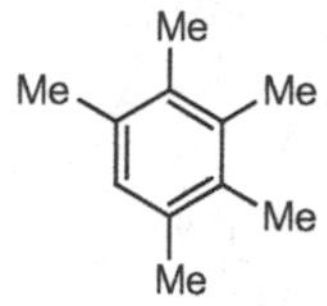	22.0
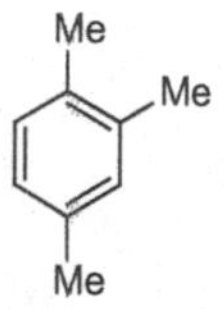	41.3	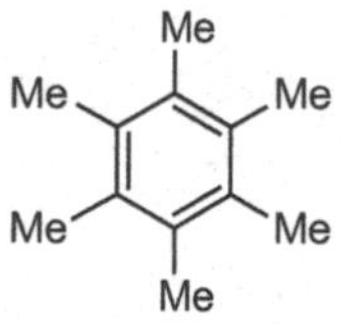	11.4

When considering the structure dependent energy of organic molecules, in this case that of substituted benzenes, all structural contributions have to be taken into account. Table 2.15 demonstrates how the H^o_{rel}-s of the substituted derivatives are additionally affected by the structure of the substituents. The effect of the structure of substituents on the energy of aromatic compounds is further supported by the data of Table 2.16. While the alkyl substituents reduce the H^o_{rel}-s compared to that of benzene, others that contain double bond, triple bond or aromatic ring considerable increase the energy.

The data of Table 2.16 show that if an unsaturated substituent is attached to the benzene ring the energy is increased. The data also offer a possibility to examine whether or not the unsaturated substituents bring about an energy reducing interaction with the aromatic benzene ring. The H^o_{rel} of styrene shows a ~62 kJ/mol increase

Table 2.15 Alkyl derivatives of benzene

Compound	H^o_{rel} (kJ/mol)	Stabilization (kJ/mol)	Compound	H^o_{rel} (kJ/mol)	Stabilization (kJ/mol)
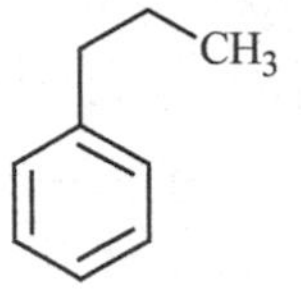 CH_3	63.0	−13.3	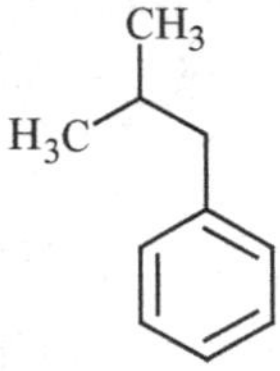 CH_3 H_3C	54.3	−22.0
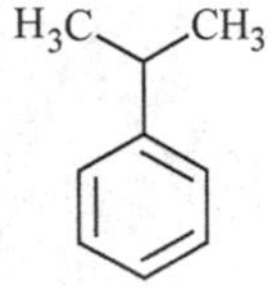 H_3C CH_3	59.1	−17.2	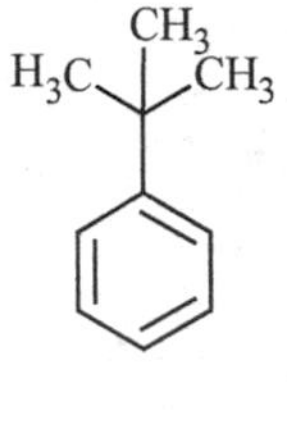 CH_3 H_3C CH_3	53.1	−23.2
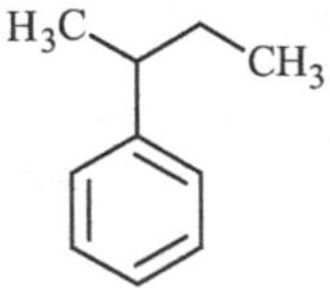 H_3C CH_3	58.4	−17.9			

Effect of the structure of substituents

relative to that of benzene. If the H^o_{rel} of styrene is compared to the sum of H^o_{rel}-s of toluene plus propene (146.3 kJ/mol) a low stabilization (7.7 kJ/mol) is observed.

This stabilization caused by the interaction of the vinyl group with the aromatic ring is relatively low compared to the 15–16 kJ/mol stabilization in butadiene. Like in the case of styrene, the stabilization in biphenyl is also weak. Its H^o_{rel} is shown to be lower by only 2.6 kJ/mol than the sum of H^o_{rel}-s of two toluenes.

$-CH_3$ H_3C-

$2 \times 64 = 128$ 125.4

Table 2.16 Benzene and naphthalene derivatives

Compound	H^o_{rel} (kJ/mol)	Effect (kJ/mol)	Compound	H^o_{rel} (kJ/mol)	Effect (kJ/mol)
	76.3			96.5	
H, H, H	138.6	62.3	CH_3	83.0	−13.5
CH_3, H, H	124.8	48.5	CH_3	82.2	−14.3
C≡C	275.0	198.7		100.8	4.3
	125.4				

If a triple bonded carbon is attached to the benzene ring no stabilization shows up at all. The sum of H^o_{rel}-s of toluene and propyne is 267.8 kJ. As it is seen the H^o_{rel} of phenylacetylene is higher (by 7.2 kJ) and not lower than this sum.

The H^o_{rel}-s of the alkyl substituted naphthalenes demonstrate that the substituents reduce the energy like the alkyl substituents of benzene do. The H^o_{rel} of acenaphthene, however, indicates a slight increase compared to that of naphthalene, instead of the expected cca. 27 kJ/mol decrease (considering the five member ring as two substituents). This is probably the consequence of the strain caused by the five member ring.

2.6.3 Polycyclic Aromatic Hydrocarbons

The H^o_{rel}-s of polycyclic aromatic hydrocarbons are demonstrated in Table 2.17. It seems reasonable to compare their H^o_{rel}-s to that of benzene. The number of constituent carbon atoms in these compounds differs considerably. For this reason comparisons are more reliable if H^o_{rel}-s are divided by the number of carbon atoms. Comparing these H^o_{rel}/n_c-s to that of benzene show that the energy per the number of carbon atoms of the polycyclic hydrocarbons (except that of biphenylene, azulene and acenaphthylene) is significantly lower than that of benzene.

In benzene all the six sp^2 carbon atoms are bonded to two other sp^2 carbon atoms. The third bond is made with hydrogen. In the fused ring polycyclic compounds, however, some of the sp^2 carbon atoms are bonded to three other sp^2 carbon atoms. This bonding structure is similar to the bonding structure of the graphite carbon atoms. So it can be supposed that some of the carbon atoms of the fused ring aromatic hydrocarbons can adopt graphite like bonding structure. H^o_{rel} of graphite is −22.85 kJ/mol (a negative value!!) so the presence of graphite-like carbon atoms in these molecules is expected to reduce the energy of these compounds. This is reflected in the lower values of H^o_{rel}/n_C-s.

It needs to be taken into account, however, that in the H^o_{rel} of graphite, in addition to bonding within the carbon sheets, the inter-sheets bonding is also reflected. So the H^o_{rel} of "gas phase graphite", or more precisely of the monolayer graphene, would certainly be somewhat higher than −22.85 kJ/mol. As far as the author is aware, however, the heat of formation of the monolayer graphene has not yet been published.

What is explained above about the role of the graphite-like carbon atoms in the fused ring polycyclic aromatic hydrocarbons is demonstrated by the data of Table 2.17. It can be seen that as the number of the only C–C bonded (n_{CC}) carbon atoms increases relative to the C–H bonding ones (n_{CH}), that is if the n_{CC}/n_{CH} increases, the H^o_{rel} per the total number of carbon atoms decreases. The data also show—as long known—that the angular arrangement of the rings is more favorable than the linear ones.

The H^o_{rel} and H^o_{rel}/n_C of biphenylene is very high as a consequence of the strained four member ring in the molecule. The second highest H^o_{rel}/n_C belongs to azulene showing that azulene is not a fully aromatic compound.

Acenaphthylene has a fully conjugated C–C bond system and in addition has four graphite-like carbon atoms, and despite this, its H^o_{rel}/n_C value is somewhat higher than that of benzene. The reason is probably the five member ring in the molecule like in the case of azulene. The n_{CC}/n_{CH} value of corannulene is the same as that of coronene. Despite this, its H^o_{rel}/n_C value is close to that of benzene, much higher than that of coronene. The reason may be again the five member ring in the molecule.

The H^o_{rel} of coronene (7.8) kJ/mol is very low. It is close to that of n-alkanes. The H^o_{rel}/n_C value is the lowest in the table. This low H^o_{rel} can be attributed to the 12

Table 2.17 Relative energy of polycyclic aromatic hydrocarbons (kJ/mol)

Compound	H^o_{rel}	H^o_{rel}/n_C	n_{CC}/n_{CH}	Compound	H^o_{rel}	H^o_{rel}/n_C	n_{CC}/n_{CH}
	76.3	12.7	0	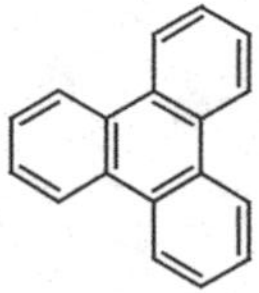	108.7	6.0	0.5
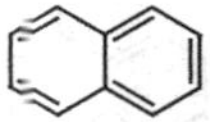	96.5	9.6	0.25		7.8	0.3	1.0
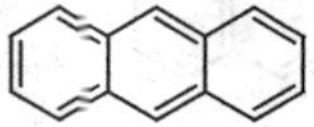	128.4	9.2	0.4	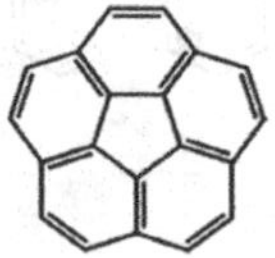	219.1	10.9	1.0
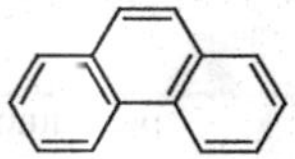	104.5	7.5	0.4	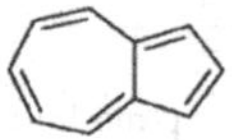	225.4	22.5	0.25

(continued)

Table 2.17 (continued)

Compound	H^o_{rel}	H^o_{rel}/n_C	n_{CC}/n_{CH}	Compound	H^o_{rel}	H^o_{rel}/n_C	n_{CC}/n_{CH}
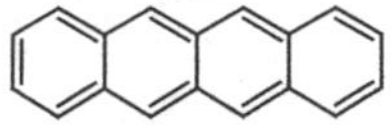	141.7	7.9	0.5	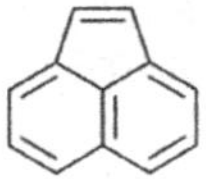	163.0	13.6	0.5
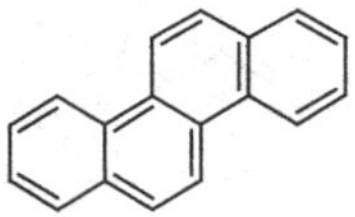	112.5	6.3	0.5	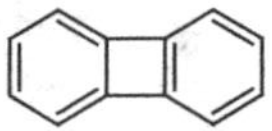	381.8	31.8	0.5
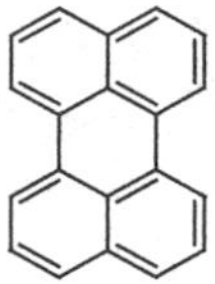	60.9	3.0	0.7	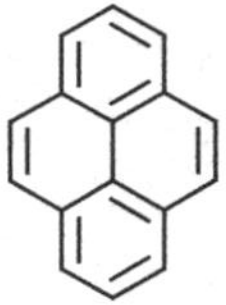	60.9	3.8	1.6

graphite-like carbon atoms of the molecule. As mentioned above, the H^o_{rel} of the graphite-like carbon is expected to be somewhat above the H^o_{rel} of elemental carbon: −22.83 kJ/mol. There are several possibilities to make rough estimations for the H^o_{rel} of graphite-like carbon. One possibility is to assign benzene CH values (that is benzene $H^o_{rel}/6 = 12.7\,\mathrm{kJ/mol}$) to the 12 CH groups of coronene and see what remains from the H^o_{rel} of coronene for each of the 12 graphite like carbon atom. The value is −12.05 kJ/mol. As expected, this is well above the H^o_{rel} of the elemental carbon. It is not justified, however, to assign H^o_{rel}-s of benzene CH groups to the H^o_{rel}-s of coronene CH groups, since the peripheral C–C bond lengths in coronene are not the same as those in benzene. Two bond lengths alternate 1.346 and 1.415 Å [5], showing that the peripheral C–C bonds have some conjugated polyene character. Assigning butadiene CH values (27.85 kJ/mol) for coronene CH groups in calculating the H^o_{rel} value of graphite-like carbon atoms, an unacceptably low value comes out: −27.2 kJ/mol. The H^o_{rel} of the graphite like carbon can't be lower, it has to be rather higher than that of the elemental carbon. There are two other possibilities for estimation. One is correction of the H^o_{rel} value of elemental carbon with the heat of sublimation of coronene per carbon atom (calculated from H^o_{rel}-s of the solid and gas phase coronene in Table 4.12) and the second, calculation of the H^o_{rel} of graphite-like carbon atom from the number of graphite like-carbon atoms, the number of CH groups and the H^o_{rel}-s of naphthalene and coronene (Table 2.17). The three realistic estimated values are:

From the H^o_{rel} value of the CH groups of benzene	$-12.05\,\mathrm{kJ/mol}$
From the H^o_{rel} value of elemental carbon corrected by the heat of sublimation of per carbon atom of coronene	$-16.88\,\mathrm{kJ/mol}$
From the $H^o_{rel}-$s, the number of graphite like carbon atoms and the number of CH groups of naphthalene and coronene	$-15.21\,\mathrm{kJ/mol}$

The best estimates are the last two values. So the approximative H^o_{rel} of the graphite carbon is expected to be around

$$-15\,\text{and}\,-17\,\mathrm{kJ/mol}.$$

References

1. Furka Á (2009) Struct Chem 20:587–604
2. Klages F (1949) Chem Ber 358
3. Dewar MLS (1969) The molecular orbital theory of organic chemistry, McGraw, New York, pp 173–177
4. Furka Á, Sebestyén F (1980) Croat Chem Acta 53:555
5. Fawcett JK, Trotter J (1966) Proc Roy Soc Lond Ser A Math Phys Sci 289(1418):366–376

Chapter 3
The Oxygen Derivatives of Hydrocarbons

3.1 Alcohols

Oxygen is a high energy element. This is shown by its high H^{o}_{rel} (125.83 kJ/mol). When it forms compounds a considerable amount of energy is evolved [1].

Alcohols can be considered to be derivatives of water. Table 3.1 shows that the H^{o}_{rel} considerably increases when one hydrogen of water is replaced by an alkyl group. Nevertheless, they are low energy compounds.

The primary alcohols derived from the reference hydrocarbons form a series with almost constant H^{o}_{rel}-s. Most of the H^{o}_{rel}-s vary around −22 and −26 kJ/mol. For the low H^{o}_{rel} of alcohols, entirely the O–H bond is responsible. Methanol, is an exception, like the first member in all other series, of derivatives of n-alkanes.

The energy of alcohols strongly depends on their order (Table 3.2). This is demonstrated by comparing the H^{o}_{rel}-s of 1-propanol (of Table 3.1), 2-propanol and tert-butanol. The stepwise increase of the order of alcohols from primary through secondary to tertiary ones the energy gradually decreases. In the H^{o}_{rel} of tert-butanol the energy reduction effects of both the 2-methyl group (8.6 kJ/mol) and the 2-hydroxyl group (52.2 kJ/mol) shows up.

CH_3-CH_2-CH_2-OH
–26.3

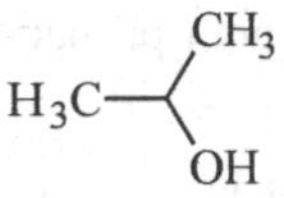

H_3C–CH(CH_3)–OH
–42.9

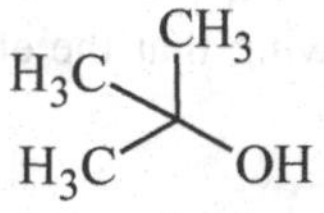

(H_3C)(H_3C)C(CH_3)–OH
–60.8

The energy reducing effect of the hydroxyl group in cyclobutanol, cyclopentanol and cyclohexanol is around −40 to −45 kJ/mol, like in the open chain secondary alkanols. The energy reducing effect of the hydroxyl group considerably exceeds that of a methyl group (~8 kJ/mol).

Á. Furka, *The Structure Dependent Energy of Organic Compounds*, SpringerBriefs in Molecular Science, https://doi.org/10.1007/978-3-030-06004-6_3

Table 3.1 Relative enthalpies of primary alcohols

Compound	H^o_{rel} (kJ/mol)	Compound	H^o_{rel} (kJ/mol)	Compound	H^o_{rel} (kJ/mol)
Water	−72.5	Heptanol	−21.2	Tetradecanol	−26.1
Methanol	−11.3	Octanol	−22.9	Pentadecanol	−26.1
Ethanol	−24.3	Nonanol	−22.1	Hexadecanol	−26.1
Propanol	−26.3	Decanol	−27.8	Heptadecanol	−26.1
Butanol	−22.6	Undecanol	−26.1	Octadecanol	−26.1
Pentanol	−25.6	Dodecanol	−26.1	Nonadecanol	−26.1
Hexanol	−26.6	Tridecanol	−26.2	Eicozanol	−26.0
Ethylene glycol	−52.9				

Table 3.2 Secondary, tertiary and unsaturated alcohols

Compound	H^o_{rel} (kJ/mol)	Effect (kJ/mol)	Compound	H^o_{rel} (kJ/mol)	Effect (kJ/mol)
2-Propanol	−42.9	−44.4	Cyclohexanol	−45.0	−45.6
2-Butanol	−40.5	−40.3	Cyclohexane	0.6	
tert-Butanol	−60.8	−52.2	Allyl alcohol	55.7	−26.6
2-Methyl-2-butanol	−56.9	−48.9	Propene	82.3	
Cyclobutanol	63.3	−45.9	Propargyl alc.	186.6	−17.2
Cyclobutane	109.2		Propyne	203.8	
Cyclopentanol	−14.1	−40.0	Propane	1.5	
Cyclopentane	25.9		Glycerol	−99.7	−101.2

If the H^o_{rel} of allyl alcohol is compared to that of propene, it turns out that the energy reducing effect of the hydroxyl group is similar to that of other primary alcohols. The same effect in propargyl alcohol is lower by ~9 kJ/mol.

Below the H^o_{rel}-s of ethylene glycol (Table 3.1) and glycerol (Table 3.2) are compared to those of 1-propanol and 2-propanol. The H^o_{rel} of ethylene glycol closely corresponds to the sum of the H^o_{rel}-s of two 1-propanol molecules (−52.6 kJ/mol) showing that the effects of both primary hydroxyl groups appear in the H^o_{rel}.

CH_3-CH_2-CH_2-OH
−26.3

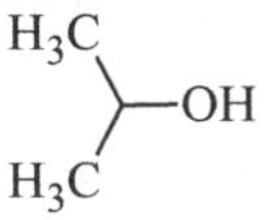

H_3C OH H_3C
−42.9

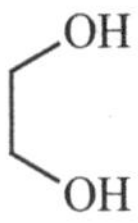

OH OH
−52.9

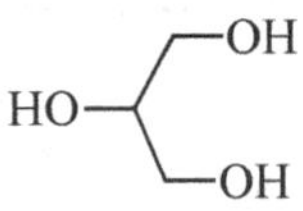

OH HO OH
−99.7

The H^o_{rel} of glycerol is close to the value expected considering its two primary and one secondary hydroxyl groups (∼−95.5 kJ/mol). The difference is only ∼4 kJ/mol.

3.1.1 Vinyl Alcohol

Vinyl alcohol is known to tautomerize to acetaldehyde. Nevertheless its gas phase heat of formation (−128 kJ/mol) has been determined experimentally. The H^o_{rel} calculated from that is 55.4 kJ/mol. This H^o_{rel} is close to the sum (50.6 kJ/mol) of the H^o_{rel}-s of ethylene and 2-propanol. The reason of tautomerization as shown below is the much lower energy of its isomeric acetaldehyde. The difference is high, 54.7 kJ/mol.

93.5 −42.9 55.4 0.7

3.2 Phenols

The presence of the hydroxyl group in phenols is expected to bring about stabilization. The question is how it compares to the stabilization of alcohols.

The data of Table 3.3 shows that the H^o_{rel} of phenol considerably differs from that of benzene. The difference is ∼53 kJ/mol! This is twice as high as the 22–26 kJ/mol stabilization in primary alcohols. In phenols the hydroxyl group is attached, however, to a tertiary carbon. So it is more appropriate to compare phenol to the tertiary alcohols. The H^o_{rel} of the tert-butanol (−60.8 kJ/mol) is lower by 52.2 kJ/mol than that of isobutane (−8.6 kJ/mol). This means that the stabilization effect of the hydroxyl group in phenols and tertiary alcohols is practically the same.

		Reduction, kJ/mol
76.3	22.8	53.5
$(CH3)_3CH$	$(CH3)_3C\text{-}OH$	Reduction, kJ/mol
−8.6	−60.8	52.2

Table 3.3 Relative enthalpy of phenols compared to that of benzene and toluene

Compound	H^o_{rel} (kJ/mol)	Effect (kJ/mol)	Compound	H^o_{rel} (kJ/mol)	Effect (kJ/mol)
	76.3		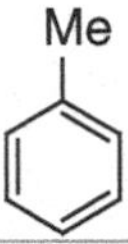 Me	64.0	
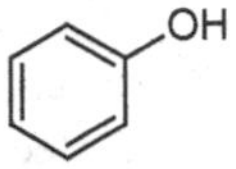 OH	22.8	−53.5	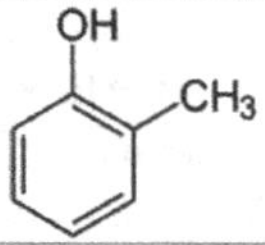 OH, CH_3	11.2	-65.1
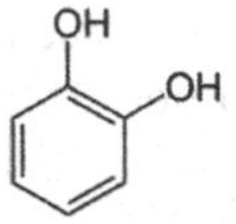 OH, OH	−29.7	−106	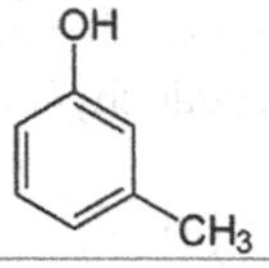 OH, CH_3	7.5	−68.8
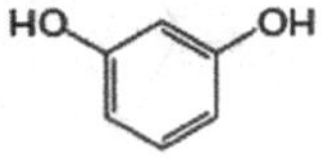 HO, OH	−39.6	−115.9	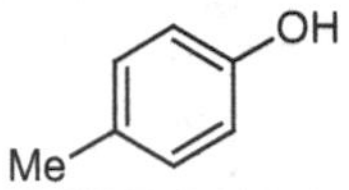 OH, Me	14.4	−61.9
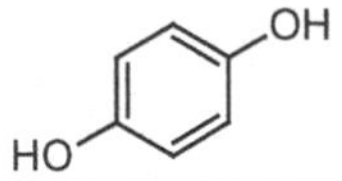 OH, HO	−31.0	−107.3			

The second hydroxyl group in dihydroxybenzenes increases stabilization. The effect of the second hydroxyl group in pyrocatechol and hydroquinone is practically the same as that of the first one: 53 kJ/mol. The same effect in resorcinol is stronger by ~10 kJ/mol. It is remarkable, that the presence of the two hydroxyl groups in dihydroxybenzenes turns benzene that has a H^o_{rel} of considerable positive value, into compounds having a remarkable negative H^o_{rel}-s.

The methyl groups in cresols bring about further stabilization. Although this stabilization somewhat differs depending on the position of the methyl substituent, it is close to the stabilization of the methyl group in toluene.

3.3 Ethers

In ethers both bonds of oxygen are made with carbon. This is why the n-alkyl ethers were chosen as reference compounds for determining the H^o_{rel} of oxygen. Table 3.4 shows that as expected for reference compounds, the H^o_{rel} of the diethyl, dipropyl and dibutyl ethers are very close to zero. The H^o_{rel} of dimethyl ether is considerably higher. This is only the manifestation of the phenomenon that is observed in all such series. If only one of the two alkyl groups is methyl group and the second is a

Table 3.4 Relative enthalpies of ethers

Compound	H^o_{rel} (kJ/mol)	Compound	H^o_{rel} (kJ/mol)
Dimethyl ether	26.4	Diisopropyl ether	−25.8
Methyl ethyl ether	14.8	Di-sec-Butyl ether	−26.5
Methyl propyl ether	14.1	Ethyl tert-butyl ether	−24.8
Diethyl ether	−0.4	Isopropyl-tert-butyl ether	−44.6
Dipropyl ether	0.1	Di-tert-butyl ether	−30.6
Dibutyl ether	0.3	Divinyl ether	152.2
Methyl isopropyl ether	−0.2	Ethyl phenyl ether	58.9

secondary one, the value of H^o_{rel} is reduced. The H^o_{rel} is further reduced if secondary alkyl groups appear in the both sides of the oxygen atom. The lowest H^o_{rel} belongs to the isopropyl tert-butyl ether and not to the di-tert-butyl ether as expected. The explanation is probably steric compression in the latter ether.

Alcohols and ethers are derivatives of water. If the hydrogens of water are replaced by one then two n-alkyl groups the energy gradually changes from the low energy of water to the zero energy of ethers.

H-O-H	CH_3CH_2-OH	CH_3CH_2-O-CH_2CH_3
−72.5	−24.3	−0.4

The H^o_{rel} of divinyl ether of course reflects the high energy of the vinyl groups. It can be compared to the H^o_{rel} of 1,4-pentadiene. The values below show that the oxygen atom is a somewhat better energy reducing component of the molecule (the difference is ~13 kJ/mol) than the methylene group.

CH_2=CH-CH_2-CH=CH_2	CH_2=CH-O-CH=CH_2
165.0	152.2

The H^o_{rel} of ethyl phenyl ether (Table 3.4) shows that the ethoxy group significantly reduces the energy as compared to that of benzene. It is a question whether or not the oxygen atom attached to the aromatic ring brings about a stronger energy reduction than in alkyl ethers. Since in ethyl phenyl ether the oxygen atom is attached to a tertiary carbon atom, its H^o_{rel} is compared below to the H^o_{rel} of ethyl tert-butyl ether. No significant difference is observed showing that there is no special energy reduction in ethyl phenyl ether. This supports the result obtained with phenols.

C_6H_6	C_6H_5-O-CH_2CH_3	Reduction kJ/mol
73.6	58.9	14.7
$(CH_3)_3CH$	$(CH_3)_3C$-O-CH_2CH_3	Reduction kJ/mol
−8.6	−24.8	16.2

3.3.1 Cyclic Ethers

Table 3.5 demonstrates that the H^o_{rel} of cyclic ethers strongly depends on ring member. The H^o_{rel} of tetrahydropyran is near to zero like that of cyclohexane. The H^o_{rel} of tetrahydrofuran is also practically the same as that of cyclopentane. The relative enthalpy of 1,4-dioxane, however, shows a slight increase compared to that of cyclohexane.

If oxygen occupies a position in a strained ring the H^o_{rel} is a high positive value. H^o_{rel}-s of ethylene oxide and of oxetane are almost the same as those of cyclopropane and cyclobutane, respectively, showing that replacement by oxygen of a CH_2 group in a three or four member ring only slightly changes the H^o_{rel}-s. This demonstrates that the accommodation of carbon and oxygen atoms to ring strain is about the same.

The H^o_{rel} of 1,3-dioxane is low compared to that of 1,4-dioxane. The reason is the acetal type bonding structure (see later).

Table 3.5 Relative enthalpies of cyclic ethers

Compound	H^o_{rel} (kJ/mol)	Compound	H^o_{rel} (kJ/mol)	Compound	H^o_{rel} (kJ/mol)
 O	114.5		115.2	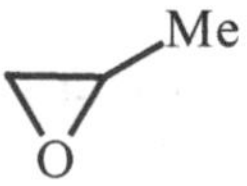 Me O	94.9
 O	107.2		109.2	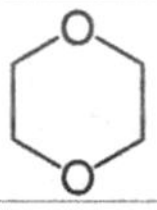 O O	18.9
 O	24.1		25.9	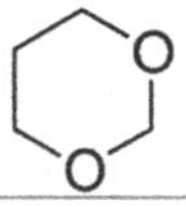 O O	−4.2
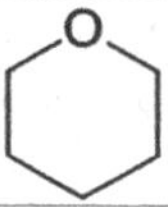 O	5.1		0.6		

3.3.2 *Stabilization Energy of Furan*

As shown before, the H^o_{rel} of the aromatic benzene (76.3 kmol) is considerably lower than that of ethylene (by 93.5 kJ/mol). Furan is also considered to be an aromatic compound. Its H^o_{rel} (93.7 kJ/mol) almost exactly equals that of ethylene. This means that the effect of the second double bond does not appear in its H^o_{rel}.

The stabilization energy can be deduced by comparing the H^o_{rel} of furan to those of cyclopentadiene and divinyl ether (Table 3.6). The H^o_{rel} of furan is lower by 56.4 kJ/mol than that of cyclopentadiene. This can be attributed in one part to replacement by oxygen of the methylene group of cyclopentadiene and in second part to the aromatic stabilization. By comparing the H^o_{rel} of divinyl ether to that of 1,4-pentadiene, the effect of CH_2 to O replacement can be deduced to be 12.8 kJ/mol. To get the stabilization energy, this value needs to be subtracted from the H^o_{rel} difference of cyclopentadiene and furan.

$$56.4 - 12.8 = 43.6$$

The result, 43.6 kJ/mol, can be considered as the aromatic stabilization energy of furan. This is much lower than the stabilization energy of benzene (90.8 kJ/mol) but is still a considerable value.

3.4 Peroxides

As reflected by its H^o_{rel} in Table 3.7, hydrogen peroxide is a high energy substance. The alkyl derivatives of hydrogen peroxide are also high energy compounds.

H^o_{rel} of dimethyl peroxide is even higher than that of hydrogen peroxide. This resembles the dimethyl ether water H^o_{rel} difference (98.9 kJ/mol) but the H^o_{rel} difference of dimethyl peroxide and hydrogen peroxide (51.9 kJ/mol) is lower. Replacement of methyl groups by ethyl groups in dimethyl peroxide reduces H^o_{rel}. A strong energy reduction relative to the H^o_{rel} of hydrogen peroxide is observable if the two hydrogen atoms are replaced by tert-butyl groups.

Table 3.6 Aromaticity of furan

Compound	H^o_{rel} (kJ/mol)	Compound	H^o_{rel} (kJ/mol)	Effect (kJ/mol)
[cyclopentadiene structure]	150.1	[furan structure] O	93.7	−56.4
CH_2=CH–CH_2–CH=CH_2	165.0	CH_2=CH–O–CH=CH_2	152.2	−12.8

Table 3.7 Relative enthalpies of organic peroxides

Compound	H^o_{rel} (kJ/mol)	Difference (kJ/mol)	Compound	H^o_{rel} (kJ/mol)	Difference (kJ/mol)
H–OO–H	159.0		tBu–OO–tBu	119.1	−39.9
Me–OO–Me	210.9	51.9	H–O–H	−72.5	
Et–OO–Et	184.7	25.7	Me–O–Me	26.4	98.9

3.5 Aldehydes and Ketones

Table 3.8 shows, the relative enthalpies of non-branched aldehydes are nearly zero like those of the reference ethers. Formaldehyde, the first member of the series, is again an exception. Its H^o_{rel} is well over zero.

Comparison of the H^o_{rel}-s of formaldehyde and carbon monoxide demonstrates that introducing two hydrogen atoms into carbon monoxide considerably increases the structure dependent energy of the molecule.

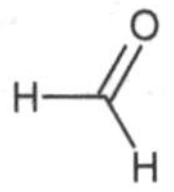

CO −7.5 30.6

Below the energy of formaldehyde is compared to that of ethylene. The comparison shows that the replacement of one of the two carbon atoms of ethylene by oxygen atom reduces the energy by 63 kJ/mol! This indicates that oxygen favors the double bond.

93.5 30.6

It has already been shown that replacing one hydrogen atom of ethylene by carbon atom (methyl group) the energy is reduced by 11 kJ/mol (see it in brackets). Doing the same with one hydrogen atom of formaldehyde the energy is reduced by a considerably higher value, 30 kJ/mol.

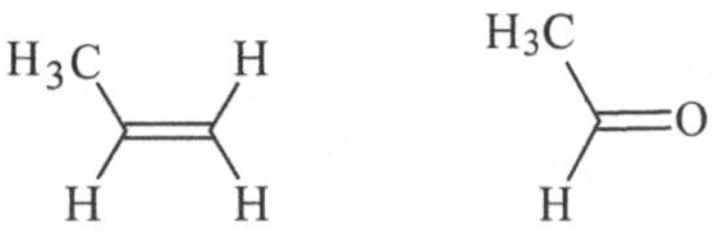

82.3 (11.2) 0,7 (29.9)

The H^o_{rel} of glyoxal is somewhat higher than that of formaldehyde but it is not twice as high. Replacement of one of its two hydrogens leads to methylglyoxal and the H^o_{rel} drops by ~38 kJ/mol to practically zero. Replacement of the second hydrogen atom decreases the H^o_{rel} by almost the same value, ~36 kJ/mol.

30.6 37.5 −0.9 −36.5

The H^o_{rel} of crotonaldehyde (55.3 kJ/mol, Table 3.8) can be compared to that of 2-butene. The geometry of crotonaldehyde is undefined and is compared to trans-2-butene, (71.3 kJ/mol). Replacement of one of the two methyl groups by formyl group reduces the energy by 16 kJ/mol.

If the methyl group of toluene is replaced by formyl group (Table 3.8) only a slight energy reducing effect is observed (~4 kJ/mol).

The H^o_{rel}-s of the non-branched ketones are lower by around 30 kJ/mol than those of the corresponding aldehydes as demonstrated in Table 3.9. The data below show that the stepwise replacement by alkyl groups of the two hydrogens in formaldehyde the energy in each step is reduced by ~30 kJ/mol.

CO −7.5 30.6 0.7 −29.9

It is interesting to note, that if carbon monoxide is transformed to formaldehyde by attachment of two hydrogens H^o_{rel} is increased by nearly 40 kJ/mol. If two methyl groups are attached, H^o_{rel} drops by around 20 kJ/mol.

Table 3.8 Aldehydes

Compound	H^o_{rel} (kJ/mol)	Compound	H^o_{rel} (kJ/mol)
CO	−7.5	Octanal	1.1
H–CHO	30.6	Nonanal	1.1
Paraformaldehyde	−21.7	Decanal	1.1
Me–CHO	0.7	OHC–CHO	37.5
Et–CHO	−4.3	Me–CO–CHO	−0.9
Pr–CHO	3.3	Me–CH=CH–CHO	55.3
Bu–CHO	1.1	Me–CH=CH–Me	71.3
Hexanal	1.2	Ph–CHO	59.6
Heptanal	6.2	Ph–Me	64.0

Table 3.9 Ketones

Compound	H^o_{rel} (kJ/mol)	Compound	H^o_{rel} (kJ/mol)
Dimethyl ketone	−39.9	2,3-butanedione	−36.5
Ethyl methyl ketone	−30.1	2,4-pentandione	−73.1
Methyl propyl ketone	−29.8	Acetophenone	30.1
Diethyl ketone	−29.8	Diphenyl ketone	96.2
Methyl isopropyl ketone	−33.7	1,2-Diphenylethane	120.1
Ethyl isopropyl ketone	−36.5	Benzyl	92.6
tert-Butyl methyl ketone	−40.3	1,4-Cyclohexadiene	139.2
Cyclohexanone	−24.0	p-Benzoquinone	83.2

If the H^o_{rel}-s of ethyl isopropyl ketone and tert-butyl methyl ketone are compared to that of acetone reveals that energy of the molecule drops if higher order carbon atom is linked to the carbonyl carbon atom. As expected, the H^o_{rel} of cyclohexanone is near to those of the open chain analogs.

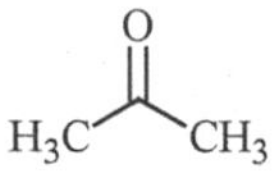

−29.9 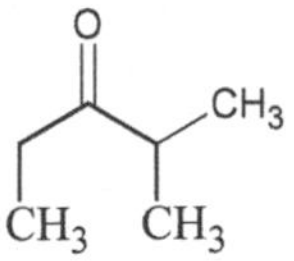 −36.5 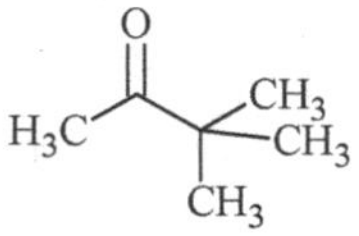 −40.3 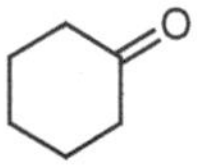 −24.0

The H^o_{rel} of acetophenon shows a ~34 kJ/mol energy reduction if compared to that of ethylbenzene. This is the same as the reduction in acetone.

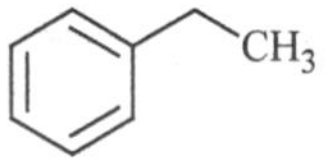

64.4 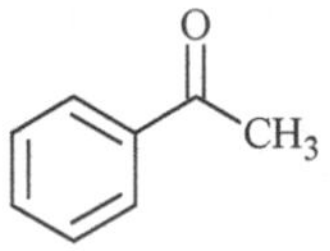 30.1

Butane-2,3-dione has two carbonyl groups. Based on this, its expected H^o_{rel} would be around −60 kJ/mol. Table 3.9 shows, however, that the H^o_{rel} is only −36.5 kJ/mol about half of the expected value. Below, this value is compared to the H^o_{rel} of 2,4-pentadione. In this molecule the two carbonyl groups are separated by a methylene group and its H^o_{rel} is even below the expected −60 kJ/mol. The comparison of the two H^o_{rel}-s clearly shows that the vicinity of the two carbonyl groups is the reason of the increased H^o_{rel} that reflects the lower stabilization in the 2,3-butanedione.

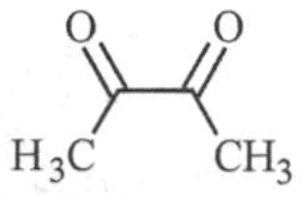

−36.5 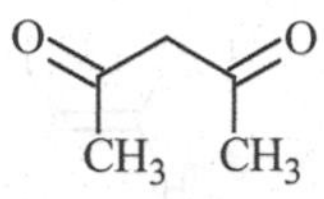 −73.1

Benzil also has two adjacent carbonyl groups that compared to 1,2-diphenylethane brings about only about 27 kJ/mol energy reduction instead of the expected around 60 kJ/mol. This again shows that the vicinity of the two carbonyl groups is unfavorable.

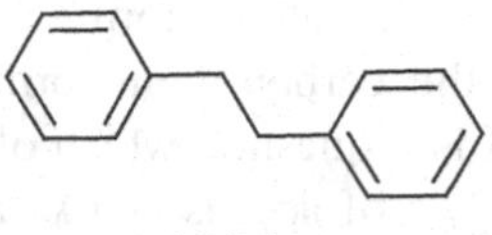

120.1 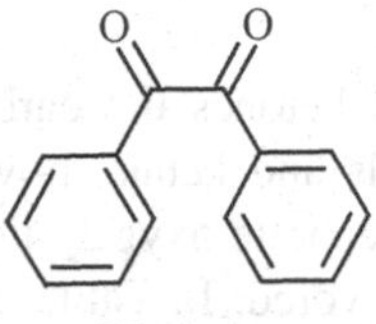 92.6

The H^o_{rel} of benzophenone can be compared to that of diphenylmethane. The difference is ~32 kJ/mol very near to the 30 kJ/mol observed in dialkyl ketones. This makes unlikely the existence of a strong energy reducing interaction between the carbonyl group and the two aromatic rings.

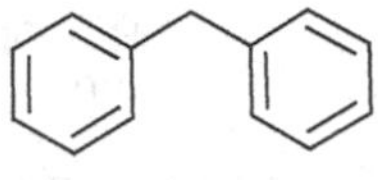

128.7 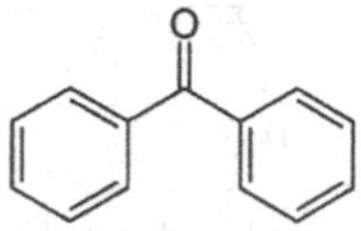 96.2

In p-benzoquinone two non-adjacent carbonyl groups are present. These two groups reduce the energy by 56 kJ/mol relative to 1,4-cyclohexadiene. This roughly corresponds to the effect of two carbonyl groups in alkanones also indicating no special interactions.

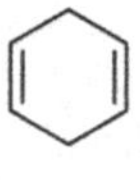

139.2 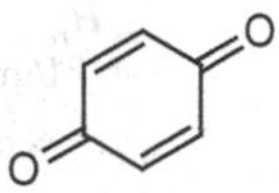 83.2

The H^o_{rel}-s of aldehydes and ketones supports an already mentioned special feature of oxygen: it better accommodates to double bonding than carbon.

This is clearly seen by comparing the H^o_{rel}-s of the pairs of compounds below. In each case a double bonded carbon atom is replaced by an oxygen atom, and a strong energy reduction is indicated by the H^o_{rel}-s of the compounds. The differences (from 62.9 to 94.4 kJ/mol) are found in the lower line. The reduced H^o_{rel}-s of the oxo compounds compared to those of the parent alkenes prove the above mentioned special feature of oxygen.

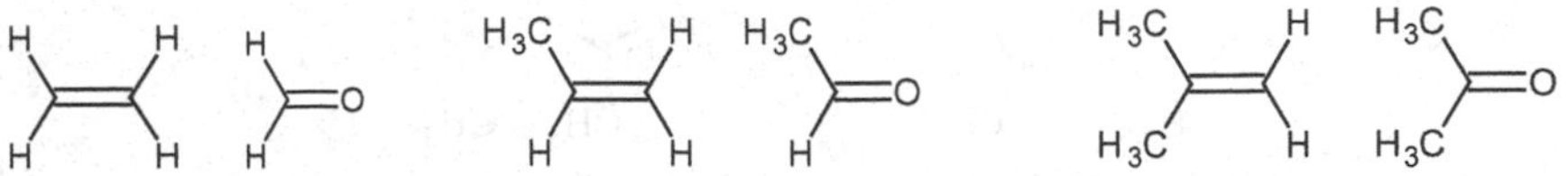

93.5 62.9 30.6 82.3 81.6 0.7 64.6 94.4 −29.9

3.6 Acetals and Ketals

In aldehydes and ketones the carbonyl carbon atoms forms a double bond with oxygen. In acetals and ketals, however, the carbon atom forms two single C–O bonds with two separate oxygen atoms. It is a question which of the two structures is energetically favored. In Table 3.10 H^o_{rel}-s of acetals and ketals are listed.

Comparing the H^o_{rel} of formaldehide dimethylacetal to that of formaldehyde shows that the H^o_{rel} acetal is more than 20 kJ/mol lower than that of formaldehyde. The difference is even higher (~51 kJ/mol) if the H^o_{rel} of formaldehyde is compared to that of its dibutylacetal.

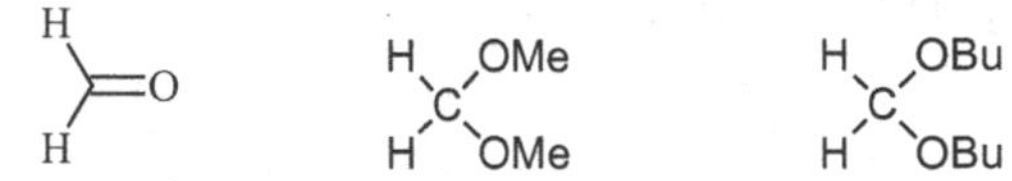

30.6 8.6 −20.5

The H^o_{rel} is reduced by about 13 kJ/mol if acetaldehyde (0.7 kJ/mol) is converted to its dimethylacetal. In the case of diethylacetal, however, the energy reduction is more significant: 35.4 kJ/mol.

Table 3.10 Relative enthalpies of acetals and ketals

Compound	H^o_{rel} (kJ/mol)	Compound	H^o_{rel} (kJ/mol)
Dimethoxymethane	8.6	Dibutoxymethane	−20.5
1,1-Dimethoxyethane	−12.6	1,1-Diethoxyethane	−34.7
2,2-Dimethoxypropane	−28.1	2,2-Diethoxypropane	−67.7
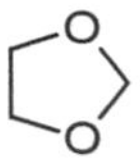	16.1	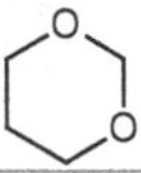	−16.0
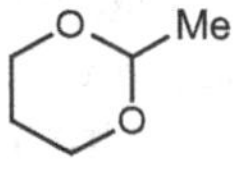	−44.2	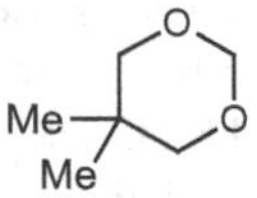	−45.8
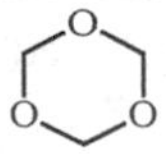	−34.2	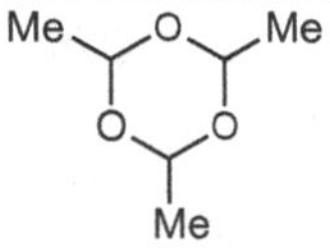	−143.9

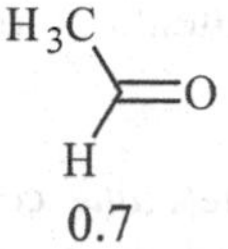

0.7 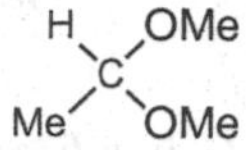 −12.6 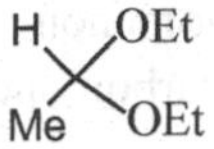 −34.7

The H^{o}_{rel} of acetone dimethylketal is close to that of acetone. The H^{o}_{rel} of acetone diethylketal, however, is considerable lower (by ~38 kJ/mol).

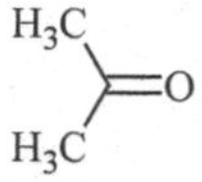

−29.9 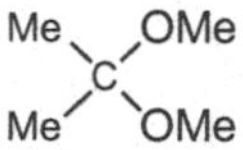 −28.1 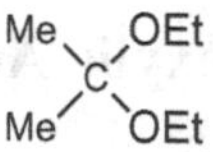 −67.7

The H^{o}_{rel}-s of the cyclic acetals listed in Table 3.10 further support the conclusion:

forming of two single C–O bonds by a carbon atom is energetically more favorable than making one double C–O bond.

The very low H^{o}_{rel} of paraldehyde is particularly convincing.

The preference of the two C–O single bonds over the one C–O double bond is probably true for hemiacetals and hemiketals, too.

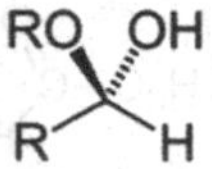

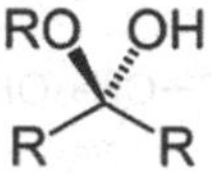

This can't be proved experimentally, since the hemiacetals and hemiketals are instable compounds and so their heats of formation are not available. The preferred cyclic structures of carbohydrates, however, are good examples to support the supposition.

The tendency that the two C–O bonds made by the same carbon atom is energetically favored continues toward the three and four such bonds. This is clearly reflected below by the H^{o}_{rel}-s of the methoxy and ethoxy derivatives of methane.

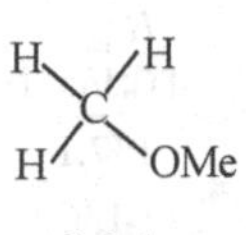

26.4 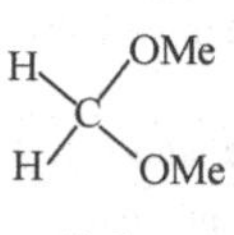 8.6 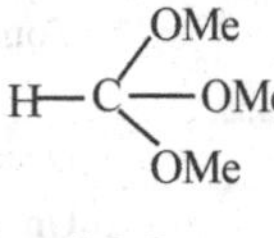 −28.3 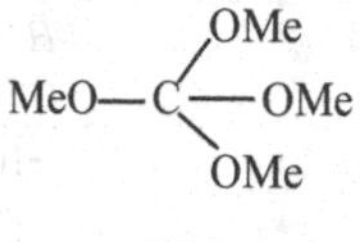 −77.3

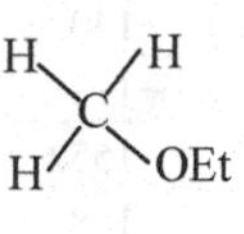

14.8 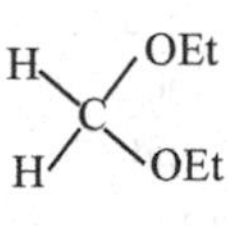 −14.9 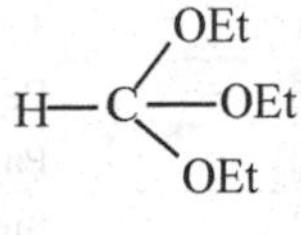 −65.3 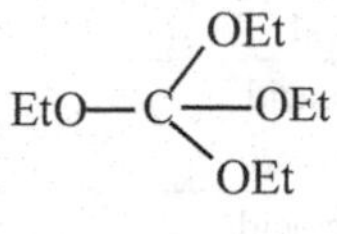 −129.3

The phenomenon manifested in these H^o_{rel}-s calls attention to a remarkable property of carbon and oxygen:

by increasing the number of C–O bonds made by the same carbon atom the energy is gradually reduced. The multiple C–O bonds made by a carbon atom are definitely favored.

3.7 Carboxylic Acids

The data listed in Table 3.11 show that carboxylic acids are low energy compounds. The relative enthalpies of the linear monocarboxylic acids are around −140 kJ/mol. The reduction of energy is the result of the hydroxyl group attached to a carbonyl group.

Comparing the data below show that replacement by hydroxyl group of a hydrogen atom in ethane reduces the energy by ~24 kJ/mol. Replacement of the carbonyl hydrogen in acetaldehyde however, results in a stabilization gain of ~140 kJ/mol! The difference is very high.

H_3C-CH_3	$H_3C-CH_2 \cdot OH$	$H_3C-C(=O)H$	$H_3C-C(=O)OH$
0.0	24.3	0.7	−139.1

Or to make another comparison: replacement of the methylene group of propane by carbonyl group reduces the H^o_{rel} by ~31 kJ/mol. Doing the same with ethyl alcohol the energy is reduced by ~115 kJ/mol!

Table 3.11 Relative enthalpies of linear monocarboxylic acids

Compound	H^o_{rel} (kJ/mol)	Compound	H^o_{rel} (kJ/mol)
Formic acid	−106.3	Decanoic acid	137.1
Acetic acid	−139.1	Undecanoic acid	−136.1
Propionic acid	−139.9	Dodecanoic acid	−142.9
Valeric acid	−136.1	Tridecanoic acid	−140.5
Pentanoic acid	−135.6	Pentadecanoic acid	−138.1
Hexanoic acid	−138.0	Palmitic acid	−155.6
Heptanoic acid	−138.3	Stearic acid	−158.4
Octanoic acid	−137.9	Nonadecanoic acid	−141.9
Nonanoic acid	−138.3	Eicosanoic acid	−148.4

Table 3.12 Relative enthalpy of dicarboxylic acids

Compound	H^o_{rel} (kJ/mol^{-1})	Compound	H^o_{rel} (kJ/mol^{-1})
HOOC–COOH	−230.8	HOOC–CH=CH–COOH *cis*	−180.5
HOOC–$(CH_2)_2$–COOH	−281.0	HOOC–CH=CH–COOH *trans*	−175.9
HOOC–$(CH_2)_4$–COOH	−281.3		

H_3C CH_3	H_3C (O) CH_3	H_3C OH	H_3C (O) OH
1.5	−29.9	−23.5	−139.1

The above comparisons allow one to conclude:

the interaction of the hydroxyl oxygen atom with the carbonyl group in carboxylic acids brings about an exceptionally strong energy reducing interaction.

In fact this interaction is the strongest seen so far.

Considering the relative enthalpies of the monocarboxylic acids listed in Table 3.11, it can be observed again, that the first member of the series differs from all others. The H^o_{rel} of formic acid is considerably higher (by ~33 kJ/mol) than those of the other members.

It also seems worthwhile to compare the structure dependent energy of the carboxylic acids to that of carbon dioxide.

CH_3-COOH	O=C=O
−139.1	−164.6

It can be seen: the H^o_{rel} of carbon dioxide is even lower than that of the low energy acetic acid. The effect of the two carbon-oxygen double bonds made by the same carbon atom on energy, can be illustrated by a second example below showing that replacement of the first two hydrogen atoms in methane by oxygen increases the energy by ~41 kJ/mol. Replacement of the second two hydrogens, however, reduces the energy by a very high value: ~195 kJ/mol!

H_2CH_2	H_2C=O	O=C=O
−10.8	30.6	−164.6

It is worthwhile to mention that the H^o_{rel} of carbon dioxide represents the lowest energy seen so far that can be attributed to a structure linked to a single carbon atom.

The data below show that replacement of a hydrogen atom in benzene by a carboxyl group reduces the energy by ~144 kJ/mol, that is close to the value observed at the saturated carboxylic acids. The difference, when ethylene and acrylic acid is compared is significantly higher (~160 kJ/mol).

C_6H_5–H	C_6H_5–COOH	$H_2C=CH_2$	$H_2C=CH$–COOH
76.3	–68.0	93.5	–66.1

Table 3.12 shows that the H^o_{rel} of dicarboxylic acids, glutaric acid and adipic acid corresponds to the sum of H^o_{rel}-s of two monocarboxylic acids.

The H^o_{rel} of oxalic acid, however, is considerably higher (by about 50 kJ/mol) than that of the above mentioned two acids, showing that the vicinity of the two carboxyl groups is energetically unfavorable.

The H^o_{rel} of fumaric acid (−180.5 kJ/mol) is nearly the sum of the effects of the three functional groups in the molecule.

$$93.5\,(\text{ethylene}) + 2\text{x}(-138)\,(\text{carboxyl group}) = -182.5\ \text{kJ/mol}$$

This is a phenomenon often observed: the H^o_{rel} of a compound sums up from the contributions linked to the individual elements of the structure.

3.8 Carboxylic Acid Esters

The H^o_{rel}-s of carboxylic acid methyl esters listed in Table 3.13 show that the esters are also low energy compounds. Their average H^o_{rel}-s is around −95 kJ/mol. The H^o_{rel}-s of ethyl, butyl and other esters are lower than that.

The following comparison shows that the low energy of esters is the consequence of the favorable interaction of the oxygen atom of the alkoxy group and the carbonyl group. If one of the methyl groups of dimethyl ketone is replaced by an ethoxy group the energy drops by ~81 kJ/mol.

H_3C–CO–CH_3	H_3C–CO–O–CH_2CH_3
−29.9	−110.7

The H^o_{rel}-s of the esters, however, are not as low as those of the carboxylic acids. This demonstrated by comparing the H^o_{rel} of ethyl acetate to that of acetic acid. The difference is around 29 kJ/mol, that can be attributed to the lack of free hydroxyl groups in esters. This is supported by the fact that the difference of the H^o_{rel}-s of diethyl ether and ethanol is ~24 kJ/mol, close to the 29 kJ/mol.

Table 3.13 Relative enthalpies of esters

Compound	H^o_{rel} (kJ/mol^{-1})	Compound	H^o_{rel} (kJ/mol^{-1})
Methyl formate	−44.0	Ethyl acetate	−110.7
Methyl acetate	−96.0	Butyl valerate	−122.9
Methyl valerate	−96.1	Isopropyl acetate	−134.9
Methyl hexanoate	−97.7	sec−Butyl butirate	−128.6
Methyl heptanoate	−100.1	Isopropyl valerate	−128.2
Methyl octanoate	−96.1	Isobutyl valerate	−131.3
Methyl nonanoate	−95.1	sec-Butyl valerate	−135.9
Methyl decanoate	−93.0	Methyl isovalerate	−122.5
Methyl undecanoate	−93.7	Ethyl isovalerate	−131.2
Methyl dodecanoate	−92.6	Methyl pivalate	−138.4
Methyl tridecanoate	−96.0	Ethyl pivalate	−140.0
Methyl tetradecanoate	−93.8		
Methyl pentadecanoate	−101.6	Allyl formate	58.9

$H_3C-C(=O)-O-CH_2-CH_3$ −110.7 $H_3C-C(=O)-OH$ −139.1

CH_3-CH_2-O-CH_2-CH_3 −0.4 CH_3-CH_2-OH −24.3

As usual, the H^o_{rel} of the first member of the series, in this case H^o_{rel} of the first member of methyl esters, differs from those of the other members. This is clearly demonstrated by comparing the H^o_{rel}-s of the first and second member.

$H-C(=O)-O-CH_3$ −44.0 $H_3C-C(=O)-O-CH_3$ −96.0

The difference is considerable: 52 kJ/mol.

As expected, and reflected by the data in the fourth column of Table 3.13, the presence of a higher order carbon atoms in the acidic or the alcoholic part of the ester reduces the energy.

It is a question whether or not the oxygen atom attached to the olefinic carbon atom in the vinyl alcohol esters has a favorable interaction with the double bond.

Comparing the data below seems to show a very low, ~6 kJ/mol stabilization. Replacements by a CH_3–COO group in ethane and ethylene show a 110.7 and a 116.4 kJ/mol reduction in the H^o_{rel}-s, respectively.

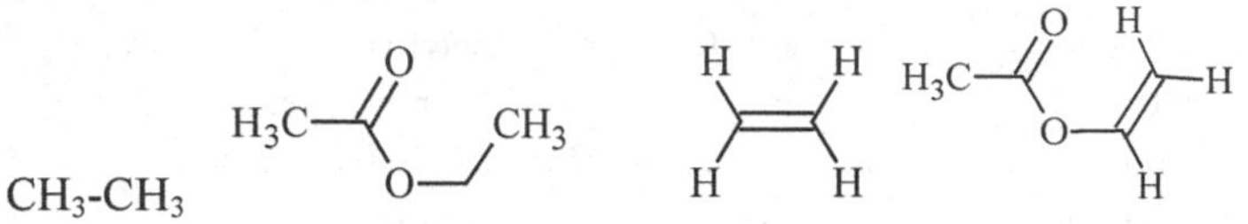

CH_3-CH_3	ethyl acetate	ethylene	vinyl acetate
0.0	−110.7	93.5	−22.9

Differences: −110.7 −116.4

It can be deduced from the H^o_{rel}-s of two molecules of benzene and that of phenyl benzoate that the COO group in the latter lowers the energy of the molecule by ~122 kJ/mol. This value is higher than the ~111 kJ/mol indicated by the H^o_{rel} comparison of ethane and ethyl acetate, and this may reflect a stabilizing interaction in phenyl benzoate.

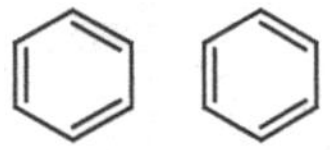

2x76.3=152.6 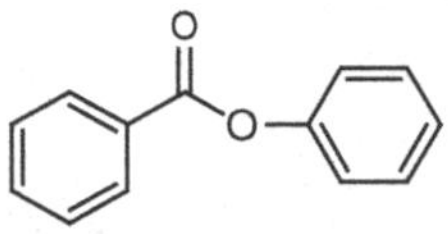 30.2

The H^o_{rel}-s of four dicarboxylic esters are found in Table 3.14. Comparing the H^o_{rel} of diethyl oxalate to the sum of H^o_{rel}-s of two ethyl acetates ($2 \times -110.7 = -221.4$ kJ/mol) indicates that bonding of the two carbonyl groups in oxalate is energetically unfavorable (by around 63 kJ/mol) like the two carboxyl groups in oxalic acid (by 50 kJ/mol).

Lactones are cyclic esters and so their H^o_{rel}-s are expected to be lower than the H^o_{rel}-s of the corresponding cycloalkanes. The data of Table 3.15 demonstrate that this is the case. The highest energy reduction belongs to propiolactone that is followed by butyrolactone then by caprolactone.

Table 3.14 Dicarboxylic acid esters

Compound	H^o_{rel} (kJ/mol)	Compound	H^o_{rel} (kJ/mol)
Dimethyl-oxalate	−166.9	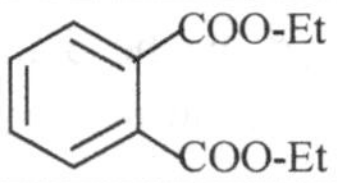 COO-Et, COO-Et	−54.7
Diethyl-oxalate	−158.3	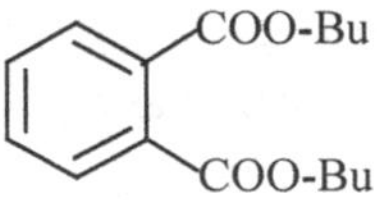 COO-Bu, COO-Bu	−70.3

Table 3.15 Comparing H^o_{rel}-s of lactones and cycloalkanes

Compound	H^o_{rel} (kJ/mol)	Compound	H^o_{rel} (kJ/mol)	Difference (kJ/mol)
	109.2	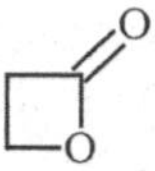 O, O	−16.1	−125.3
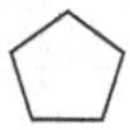	25.9	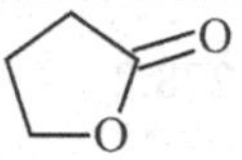 O, O	−73.3	−99.2
	25.0	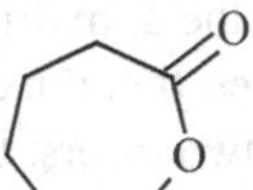 O, O	−64.0	−89.0

3.9 Carboxylic Acid Anhydrides

As it is well known carboxylic acid anhydrides are reactive reagents. Their H^o_{rel}-s listed in Table 3.16, however, show that they are low energy compounds. The low energy of anhydrides can be attributed to the fact that their carbonyl groups are bonded to oxygen like in acids and esters.

In order to show the stabilization effect in the three groups of compounds the H^o_{rel} of acetic anhydride is compared to those of acetic acid and ethyl acetate.

Table 3.16 Carboxylic anhydrides

Compound	H^o_{rel} (kJ/mol)	Compound	H^o_{rel} (kJ/mol)
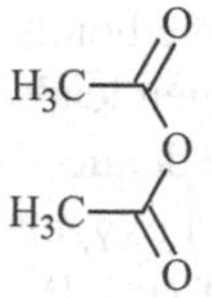 H_3C, O, O, H_3C, O	−156.9	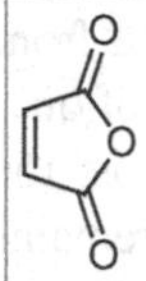 O, O, O	−68.6
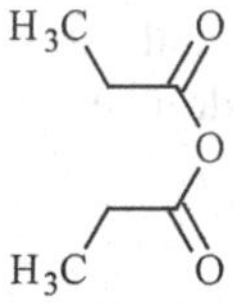 H_3C, O, O, H_3C, O	−168.7	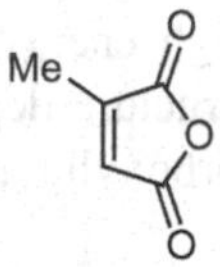 Me, O, O, O	−96.9
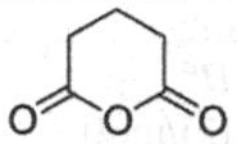 O, O, O	−138.7	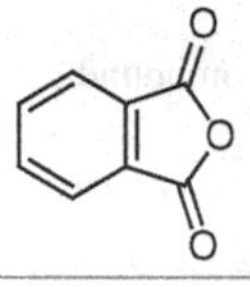 O, O, O	−89.6

$(CH_3CO)_2O$	CH_3COOH	$CH_3COOCH_2CH_3$
−156.9	−139.1	−110.7
	2x−139.1 = −278.2	2x−110.7 = −221.4

The comparison reveals that the H^o_{rel} of the anhydride is lower than those of both the acid and the ester. Considering, however, that it has two carbonyl groups its H^o_{rel} can be compared to those of two acids or two esters, too. It can be seen that H^o_{rel} of the anhydride is considerably higher than the doubled H^o_{rel}-s of acid and ester. There is good reason to believe that the less than double stabilization in the anhydride is the consequence of the fact that a single oxygen atom interacts with two carbonyl groups. This is probably less effective than if two oxygen atoms interact with two carbonyl groups.

If the H^o_{rel} of maleic anhydride is compared to that of ethylene (93.5 kJ/mol) it can be seen, that the presence of the anhydride group reduces the H^o_{rel} of the parent molecule very considerably (by more than 160 kJ/mol). The result is about the same if the H^o_{rel}-s of benzene (76.3 kJ/mol) and that of phthalic anhydride are compared. A methyl substituent in maleic anhydride reduces the H^o_{rel} by nearly 30 kJ/mol.

3.10 Ketene and Carbon Dioxide

Ketene is a reactive compound and its H^o_{rel} is a considerable positive value (Table 3.17). Ketene can be deduced from allene that has cumulated double bonds and a high positive H^o_{rel}. Replacing of one of its methylene groups by oxygen leads to ketene and H^o_{rel} is reduced by almost 150 kJ/mol. Carbon dioxide can be deduced from ketene by replacement of the remaining methylene group by a second oxygen atom. This brings about an even higher—nearly 230 kJ/mol—reduction of H^o_{rel}. By these two replacements the high energy allene changes to the exceptionally low energy carbon dioxide. The structure dependent energy of carbon dioxide is even lower than that of the monocarboxylic acids.

Table 3.17 Ketene and carbon dioxide

Compound	H^o_{rel} (kJ/mol)	Compound	H^o_{rel} (kJ/mol)
$CH_2{=}C{=}CH_2$	210.5	$CH_2{=}C{=}O$	62.5
$O{=}C{=}O$	−164.6		

The high energy reductions brought about in both stages of the replacement of the methylene groups of allene to form carbon dioxide, further support the special feature of oxygen already mentioned at the oxo compounds: oxygen accommodates better to double bonding than carbon.

Reference

1. Furka, Á.: Struct Chem **20**:587–604 (2009)

Chapter 4
Organonitrogen Compounds

4.1 Amines

Triethylamine is the reference substance for nitrogen compounds so the H^o_{rel} of it is assigned to be zero. As reflected by the H^o_{rel} in Table 4.1, nitrogen is an exceptionally low energy element. For this reason all of its derivatives listed in the table have much higher H^o_{rel}-s. Although the H^o_{rel} of its simplest derivative, ammonia, has a considerably high negative value, it is far from that of nitrogen [1].

Amines can be considered to be derivatives of ammonia. It is shown below that as the hydrogen atoms of ammonia are replaced by n-alkyl groups, the H^o_{rel} gradually increases up to zero. The reason is the reduction of the number then the disappearance of the N–H bonds from the molecules.

NH_3	CH_3CH_2-NH_2	CH_3CH_2-NH-CH_2-CH_3	CH_3CH_2-N-CH_2-CH_3 1 CH_2CH_3
–70.1	–28.9	–14.1	0.0

This means that the relative energy of amines depends on their order:

$$\text{primary} < \text{secondary} < \text{tertiary}$$

The result of the above comparisons shows that the N–H bonds are energetically more favorable than the N–C bonds. In this respect nitrogen is similar to oxygen: the O–H bonds are more favorable than the O–C bonds.

The energy represented by the N–H bonds can be compared to those of the O–H bonds, too.

Á. Furka, *The Structure Dependent Energy of Organic Compounds*, SpringerBriefs in Molecular Science, https://doi.org/10.1007/978-3-030-06004-6_4

Table 4.1 Amines

Compound	H^o_{rel} kJ/mol	Compound	H^o_{rel} kJ/mol
Nitrogen (N_2)	−178.6	sec-Butylamine	−45.9
Ammonia	−70.1	tert-Butylamine	−61.6
Methylamine	−26.5	Dimethylamine	−1.4
Ethylamine	−28.9	Diethylamine	−14.1
Propylamine	−34.7	Dipropylamine	−16.9
Butylamine	−33.7	Dibutylamine	−30.3
Hexylamine	−94.3	Butyl-isobutylamine	−34.1
Heptylamine	−95.8	Di-sec-butylamine	−42.8
Isopropylamine	−46.1	Trimethylamine	13.9
Isobutylamine	−40.3	Triethylamine	0.0
Isopentylamine	−116.9	Tripropylamine	0.6
Cyclopropylamine	71.3	Ethylenediamine	−67.5
Allylamine	23.1		

$(CH_3CH_2)_2N - H$ $\quad$ $CH_3CH_2 - O - H$
−14.1 $\quad$ −24.3

Comparison of the H^o_{rel}-s of diethylamine and ethanol reveals that the hydroxyl group of alcohols ensures a lower energy for the molecule than does the N–H bond for the secondary amines. The N–H bond is energetically less favorable than the O–H bond.

The H^o_{rel} of amines also strongly depends on the order of the carbon atom that makes the bond with the nitrogen of the amino group. This is exemplified below and is compared to the dependence of the H^o_{rel} of alcohols on their order.

$CH_3\text{-}CH_2\text{-}NH_2$ −28.9 $\quad$ H_3C–CH(CH_3)–NH_2 −45.9 $\quad$ $(H_3C)_2C(CH_3)NH_2$ −61.6

$CH_3\text{-}CH_2\text{-}OH$ −24.3 $\quad$ H_3C–CH(CH_3)–OH −42.9 $\quad$ $(H_3C)_2C(CH_3)OH$ −60.8

Increasing the order of the alkyl group, the H^o_{rel}-s of amines like those of alcohols are strongly decreasing. The comparisons reveal that the dependence of the H^o_{rel} of amines and alcohols on their order is very similar.

The H^o_{rel}-s of hexylamine, heptylamine and isopentylamine in Table 4.1 differ very much from the others. One can't exclude that these discrepancies are consequences of experimental errors in the heats of formation. Other data reflect that branching in the alkyl groups also decreases the value of the H^o_{rel}-s.

The H^o_{rel} of ethylenediamine closely corresponds to the sum of those of two alkylamines.

4.1.1 Azacycloalkanes

Table 4.2 shows the H^o_{rel}-s of a few azacycloalkanes. The H^o_{rel} of aziridine is a high positive value due to the strain in the three member ring. It is lower, however, than the H^o_{rel} of cyclopropane by ~23 kJ/mol that can be at least in part the consequence of the fact that aziridine is a secondary amine. The H^o_{rel}-s of open chain secondary amines are around −14, −30 kJ/mol. Inclusion of nitrogen into the five and six member hydrocarbon rings also reduces the H^o_{rel}-s like in the open chain secondary amines. Because of the two nitrogen atoms in the ring, this value is doubled in the case of piperazine.

Triethylenediamine is a bicyclic tertiary diamine, so its H^o_{rel} is expected to be around zero. The −7.2 kJ/mol is close to it. Hexamethylenetetramine is a tricyclic tertiary tetramine so its H^o_{rel} is also expected to be close to zero. The −34.8 kJ/mol seems to be a too low value for its H^o_{rel}. It has to be taken into account, however, that in this molecule each carbon atom is bonded to two nitrogen atoms. This is similar to bonding of two oxygen atoms to one carbon atom for example in formaldehide dimethylacetal that reduces the energy.

Table 4.2 Cyclic amines

Compound	H^o_{rel} kJ/mol	Compound	H^o_{rel} kJ/mol	Effect kJ/mol	Compound	H^o_{rel} kJ/mol
	115.2	 NH	91.9	−23.3	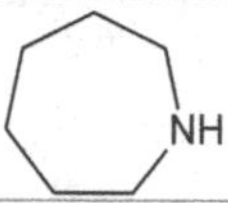 NH	−11.1
	25.9	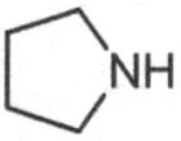 NH	11.5	−14.4	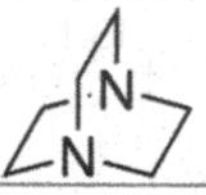 N N	−7.2
	0.6	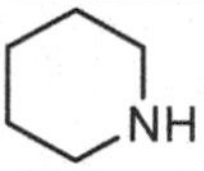 NH	−11.6	−12.2	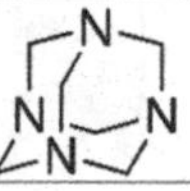 N N N N	−34.5
	0.6	 HN NH	−27.7	−28.3		

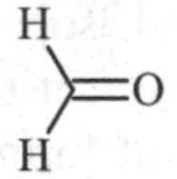

30.6 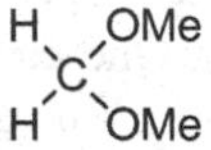 8.6

If nitrogen has a property similar to that of oxygen reflected in the reduced H^o_{rel}-s of the acetals, this may explain the low H^o_{rel} of hexamethylenetetramine. In fact the low H^o_{rel} of hexamethylenetetramine may be considered as an indication of such property of nitrogen.

4.1.2 Aromatic Amines

The H^o_{rel}-s of several aromatic amines are found in Table 4.3. The H^o_{rel} of aniline looks very low if compared to that of benzene. By substituting the hydrogen atoms of the amino group by alkyl groups the H^o_{rel} increases as expected. Nevertheless the H^o_{rel} of the N,N-diethylaniline is still considerably lower than that of benzene.

Below, the energy reducing effects of diethylamino and ethoxy substituents of benzene are compared. The comparison shows that the diethylamino group is more effective in reducing the energy. The difference looks significant, ~10 kJ/mol.

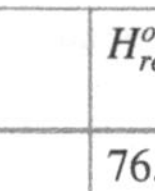

76.3 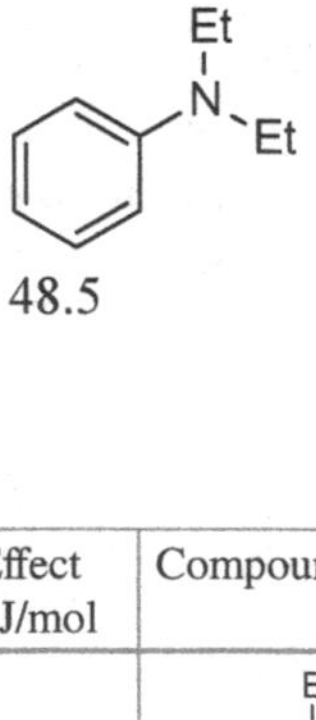 48.5 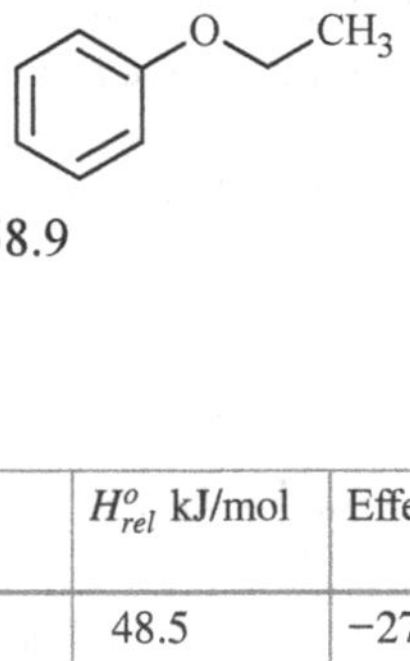 58.9

Table 4.3 Aromatic amines

Compound	H^o_{rel} kJ/mol	Effect kJ/mol	Compound	H^o_{rel} kJ/mol	Effect kJ/mol
	76.3		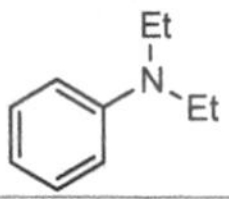	48.5	−27.8
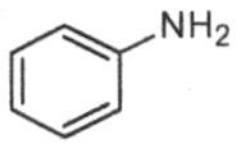	13.1	−63.2	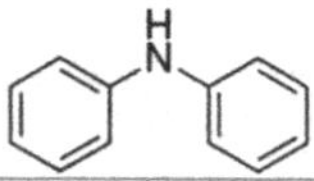	77.4	−75.2
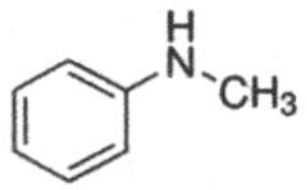	31.8	−44.5	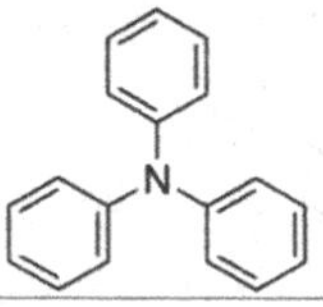	152.6	−76.3
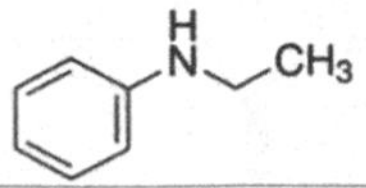	23.1	−53.2			

The H^o_{rel}-s of triphenylamine and triphenylmethane are compared below. The comparison shows that the nitrogen atom in triphenylamine more effectively reduces the energy than the central carbon atom in triphenylmethane. The difference is considerable: ~37 kJ/mol.

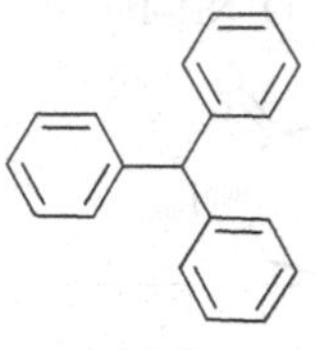

N

189.9 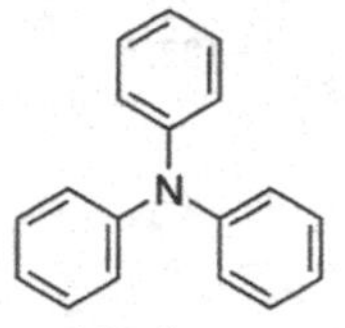 152.6

Previous comparisons demonstrated that the potential interaction of the non-bonded electron pairs of the oxygen atom with the aromatic electron system of the benzene ring in phenols and phenol ethers does not lead to decrease of energy. In the comparisons below, the effect of the electron pair of nitrogen is estimated. The comparisons below show that the amino group in aniline reduces the energy by 63.2 kJ/mol. The energy reduction effect of the amino group in tert-butylamine is somewhat lower, 53 kJ/mol. The difference, ~10 kJ/mol can be considered as the result of the extra stabilization that can be attributed to the interaction of non-bonded electron pair of nitrogen with the electrons of the benzene ring. The same result was obtained above when the effects of dimethylamino and ethoxy groups were compared.

NH_2

76.3 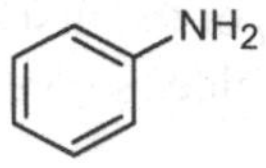 13.1 Reduction 63.2 kJ/mol

$(CH_3)_3CH$ $(CH_3)_3C\text{-}NH_2$

–8.6 –61.6 Reduction 53.0 kJ/mol

4.2 Carboxylic Acid Amides

The carboxylic acid amides are low energy compounds as demonstrated by their H^o_{rel}-s listed in Table 4.4. In this respect the amides resemble the carboxylic acid esters as exemplified by comparing below the H^o_{rel} of N,N-diethylacetamide to that of ethyl acetate.

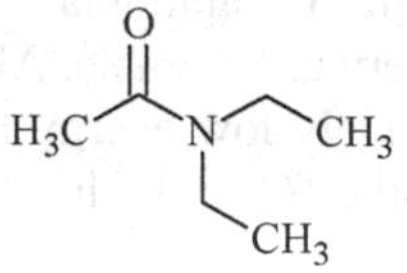

O O

H_3C N CH_3 H_3C O CH_3

CH_3

–105.2 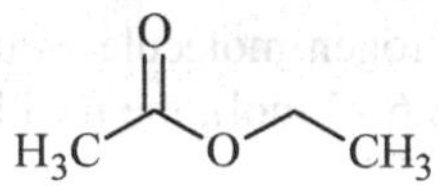 –110.8

Table 4.4 Relative enthalpies of carboxylic acid amides

Compound	H^{o}_{rel} kJ/mol	Compound	H^{o}_{rel} kJ/mol
H–CO–NH_2	−107.3	H–CO–$N(CH_3)_2$	−71.5
Me–CO–NH_2	−138.8	Me–CO–$N(C_2H_5)_2$	−105.2
Bu–CO–NH_2	−128.9	Me–CO–NH–Ph	−79.5
Me–CO–NH–Bu	−122.7	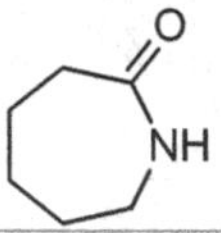 O NH	−107.5

The energy of amides, non-substituted or mono-substituted on their nitrogen atom, is lower than that of the N-disubstituted ones, as a consequence of the two and one N–H bonds that reduce the energy.

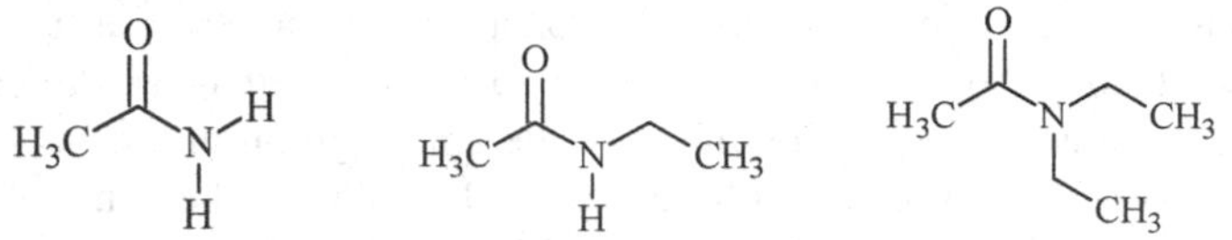

O H H_3C N H
−138.8

O H_3C N CH_3 H
−122.7

O H_3C N CH_3 CH_3
−105.2

Data of Table 4.4 also calls attention to the previously regularly observed fact that in the compound series: the H^{o}_{rel} of the first member differs from that of the rest of the members. The H^{o}_{rel} of formamide is considerably higher than those of acetamide and valeramide.

Lactames like ε-caprolactame, are cyclic amides. The H^{o}_{rel} of caprolactame can be compared to that of N-butylacetamide. The H^{o}_{rel} of the former is higher by ~15 kJ/mol that can be attributed to its seven member cyclic structure. It seems worthwhile to remind that the H^{o}_{rel} difference between the tension free cyclohexane (0.6 kJ/mol) and the seven member cycloheptane (25 kJ/mol) also exists and that is even somewhat higher than 15 kJ/mol.

4.3 Hydrazine Derivatives

Comparing the H^{o}_{rel} of hydrazine (Table 4.5) to the sum of H^{o}_{rel}-s of two molecules of ethylamine (−57.8 kJ/mol) or to two molecules of ammonia (−139.6 kJ/mol) it can be seen that the N–N bond in hydrazine is very unfavorable. Although the triple bond in the nitrogen molecule assures an extremely low energy for the elemental nitrogen (−178.6 kJ/mol), the positive value of the H^{o}_{rel} of hydrazine reflects a very week single N–N bond.

Table 4.5 Hydrazine and its derivatives

Compound	H^o_{rel} kJ/mol	Compound	H^o_{rel} kJ/mol
NH_3	−69.8	CH_3–CH_2–NH_2	−28.9
H_2N–NH_2	3.5	Me–NH–NH–Me	39.5
Me–NH–NH_2	14.3	Ph–NH–NH_2	62.0
Me_2N–NH_2	34.4		

Substitution of hydrogen atoms of hydrazine by methyl groups gradually increases the H^o_{rel}-s.

H_2N–NH_2 3.5; H_3C–NH–NH_2 14.3; $(H_3C)_2N$–NH_2 34.4; H_3C–NH–NH–CH_3 39.5

4.4 Nitriles

Nitriles are derivatives of hydrogen cyanide. Hydrogen cyanide itself can be considered as a hybrid of acetylene and nitrogen molecules.

HC≡CH	HC≡N	N≡N
224.5	40.1	−178.6

Acetylene is a very high energy compound and nitrogen is a very low energy element. Hydrogen cyanide is about half way between the two, somewhat closer to acetylene. The high energy of acetylene is a result of the unfavorable C–C triple bond. What is bad for carbon is the best for nitrogen. Since HCN is half acetylene and half nitrogen the 40 kJ/mol value for its H^o_{rel} seems reasonable.

Table 4.6 demonstrates the H^o_{rel}-s of several nitriles. Replacement by alkyl groups of the hydrogen atom in hydrogen cyanide considerably reduces the energy. In the low H^o_{rel} of isopropyl cyanide even the effect of branching in the alkyl group is observable.

The higher positive value of the H^o_{rel} of benzonitrile is understandable taking into account the effect of the benzene ring on the energy of the molecule. Comparisons below show that replacement of a hydrogen atom of ethane or benzene by CN group does not change significantly the H^o_{rel}-s.

Table 4.6 Relative enthalpies of nitriles

Compound	H^o_{rel} kJ/mol	Effect kJ/mol	Compound	H^o_{rel} kJ/mol	Effect kJ/mol
H–CN	40.1			115.2	
Me–CN	18.1	−22.0	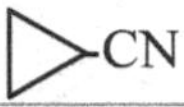 CN	127	86.9
Et–CN	1.4	−38.7	HC≡CH	224.5	219.3
Pr–CN	5.5	−34.6	 HC≡—≡N	240.8	
iPr–CN	−3.2	−43.3	NC-C≡C-CN	259.4	
Ph–CN	78.3	38.2	NC–CN	84.8	

$H_3C—CH_3$ 0.0 H_3C ≡N 1.4 76.3 ≡N 78.3

Comparing the H^o_{rel}-s of acetylene, cyanoacetylene and dicyanoacetylene shows, however, that replacement of the first then the second hydrogen atom of acetylene by cyano group in each stage increases the energy by 16–18 kJ/mol.

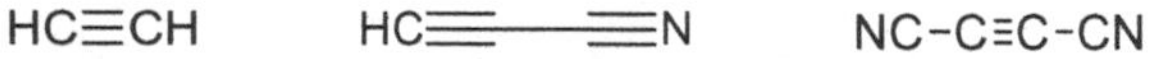

HC≡CH 224.5 HC≡—≡N 240.8 NC-C≡C-CN 259.4

Bonding of two cyano groups in cyanogen, as reflected by its H^o_{rel} is unfavorable. The H^o_{rel} roughly corresponds to that of two molecules of hydrogen cyanide.

4.5 Imines, Oximes, Hydroxylamines and Azo Compounds

As already mentioned, the double C-C bonds in hydrocarbons are sources of high energy. The C–N double bond is different. This is demonstrated below by the H^o_{rel} of acetaldimine. Despite of the double bond, the H^o_{rel} of the compound is practically zero. The low H^o_{rel} of N-benzylideneaniline compared to that of stilbene also supports the low energy of the C=N bond of imines.

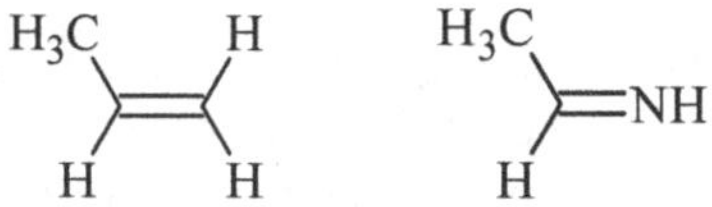

H_3C H H H 82.3 H_3C =NH H −2.3

-CH=CH- 184 (estimated) -CH=N- 131.6

The H^o_{rel} of oximes, that also have C–N double bonds, is high compared to that of imines. This is demonstrated by comparing the H^o_{rel}-s of acetaldoxime to that of acetaldimine.

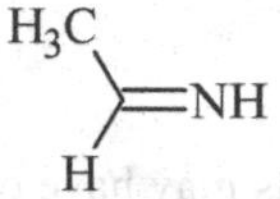

H_3C =NH H H_3C =N H OH

−2.3 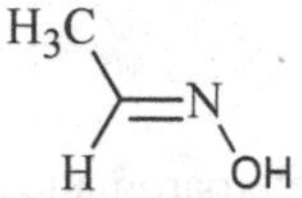   77.0

The H^o_{rel} of acetaldoxime is high compared to the lower than zero energy of acetaldimine. The reason is the hydroxyl group present in acetaldoxime. In alcohols the hydroxyl group significantly reduces the energy. In case of oximes, however, the energy is considerably higher attributable to the energetically unfavorable single N–O bond.

The unfavorable effect of the N–O bond on energy is further supported by comparing the H^o_{rel} of N,N-diethylhydroxylamine to that of diethylamine.

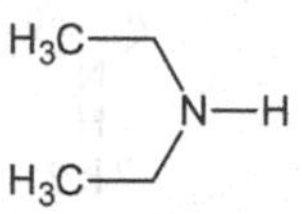

H_3C N—H H_3C H_3C N—OH H_3C

−14.1 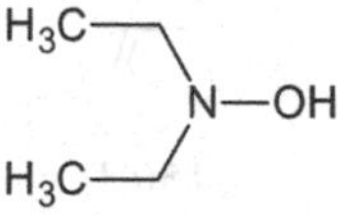   39.6

The azo compounds are derivatives of diazene, H–N=N–H that has N=N bond. The azo compounds are represented by two examples, trans-azopropane and azobutane. Their H^o_{rel}-s are compared to that of trans-2-butene.

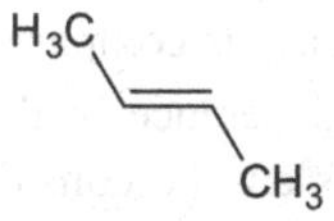

H_3C CH_3 H_3C N=N CH_3 C_4H_9-N=N-C_4H_9

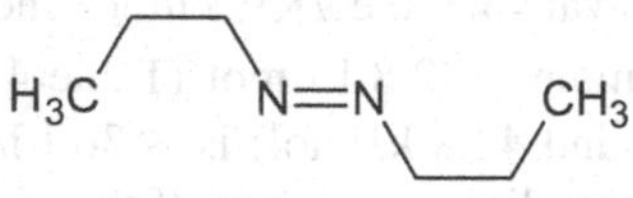

71.3   51   49.1

The H^o_{rel} of both azo-compounds is lower than that of 2-butene showing that the N=N bond in the two molecules brings about lower energy increase than the C=C bond in trans-2-butene. The energy difference between the alkene and the azo-compound is around 20–22 kJ/mol. Or in other words,

the N=N bond is less unfavorable than the C=C bond.

4.6 Heteroaromatic Compounds

4.6.1 Five Member Ring Compounds. Stabilization Energy of Pyrrole

Nitrogen containing heteroaromatic compounds may have either five or six member rings. Several five member ring representatives and their H^o_{rel}-s are listed in Table 4.7.

Pyrrole is considered to be an aromatic compound. Its H^o_{rel} can be compared to that of cyclopentadiene (151.1 kJ/mol). If the methylene group of cyclopentadiene is replaced by nitrogen, a considerable drop in H^o_{rel} is observed. Two very different H^o_{rel} values of pyrrole are seen in the table, calculated from two different experimental heats of formation. One of them was published in 1967 and the other in 1991. The newer value is used in comparisons.

The above comparison shows that replacement of the CH_2 group by NH group brings about a 78.9 kJ/mol decrease of energy. It is clear that total of this energy can't be considered as stabilization energy because the imino group itself also reduces the H^o_{rel} like in dialkyl amines. The best would be to compare the energy decrease to the difference of H^o_{rel}-s of divinylamine and 1,4-pentadiene. Since the ΔH^o_f of divinylamine is not available, the 78.9 kJ/mol energy decrease is compared to the H^o_{rel} of di-sec-butylamine, −42.8 kJ/mol (Table 4.1). The difference in the two energy decreases (78.9 and 42.8 kJ/mol) is ~36 kJ/mol. This can be considered as a value close to the stabilization energy. If the older and lower value of the

Table 4.7 Relative enthalpies of five member ring heteroaromatic compounds

Compound	H^o_{rel} kJ/mol	Compound	H^o_{rel} kJ/mol	Compound	H^o_{rel} kJ/mol
N H	36.3[1967]	N H	71.2[1991]	O	93.7
N N H	21.0	N N N H	−55.7	N O	17.7
N N H	−32.6	N–N N N H	−2.1	N O	115.2

H^o_{rel} of pyrrole is used in calculation, a 71 kJ/mol comes out for the stabilization energy. Just to remind, stabilization energy calculated for furan (43.6 kJ/mol) is between these two values.

Replacing of a CH group in the pyrrole ring by a second nitrogen atom in position 3, a very strong reduction of energy (from 71.2 to −32.6 kJ/mol) is observed. This energy drop (~104 kJ/mol) is considerably higher than that following the replacement of the methylene group in cyclopentadiene (78.9 kJ/mol). The CH substitution by nitrogen in position 2 also reduces the energy but by a lower value (from 71.2 to 21.0 kJ/mol). Substitution by two nitrogen atoms in positions 2 and 4 of pyrrole results in a very strong energy decrease: ~127 kJ/mol! This clearly demonstrates that in the five member ring, nitrogen better accommodates multiple bonds than carbon.

It is interesting to note that even four carbon atoms of the cyclopentadiene ring can be replaced by nitrogen atoms. The H^o_{rel} of tetrazole is still a negative value: −2.1 kJ/mol, although the H^o_{rel} relative to that of 1,2,4-triazole is increased by ~54 kJ/mol.

Replacing by nitrogen a CH group in position 3 in furan also results in a strong energy reduction, although its value (76 kJ/mol) is lower than that of the similar replacement in pyrrole. Substitution by nitrogen in position 2, however, increases the energy as a consequence of the unfavorable N–O bond.

4.6.2 Benzenoid Heteroaromatic Compounds

Table 4.8 shows the relative enthalpies of nitrogen containing benzenoid heterocyclic compounds. These data demonstrate the effect on energy of substitutions by nitrogen of the CH groups in benzene ring.

The H^o_{rel} of pyridine is lower than that of benzene by ~31 kJ/mol (Table 4.8). This shows that nitrogen is well accommodated in the aromatic ring with remarkable reduction of energy. The H^o_{rel}-s of pyrimidine and pyrazine demonstrate that introduction of a second nitrogen atom into the ring in non–vicinal positions, reduces the energy by a further ~32 kJ/mol. These two nitrogen substitutions reduce the energy from a moderately high value of benzene close to the level of n-alkanes.

Three nitrogen atoms in the ring in positions 1,3,5 further reduce the energy relative to pyrimidine by a very high value: ~59 kJ/mol. It is very remarkable that as a result of the triple nitrogen substitutions in the benzene ring the energy drops from the moderately high level of benzene to a moderately low value. The energy difference between benzene and 1,3,5-triazine is ~122 kJ/mol!

Replacement of two vicinal CH groups by nitrogen atoms in benzene, however, increases the energy relative to benzene by ~19 kJ/mol. Bonding of the two nitrogen atoms in pyridazine is unfavorable. It seems worthwhile to remind that a similar replacement in cyclopentadiene reduces the energy.

Replacing one CH group of naphthalene by nitrogen the energy drops to 57.8 kJ/mol. The H^{o}_{rel} of quinoline is lower by 38.7 kJ/mol than that of naphthalene. This energy reduction is somewhat even higher than that of the corresponding substitution in benzene leading to pyridine.

Methyl substitution in pyridine decreases the energy by values similar to that of the methyl group in toluene. The most preferred substitution position is vicinal to the nitrogen.

Summarizing the result obtained by analyzing the H^{o}_{rel}-s of nitriles, imines and heteroaromatic compounds reveals

a special feature of nitrogen: it better accommodates to multiple bonding than carbon.

This is illustrated by comparing below the H^{o}_{rel}-s of pairs compounds containing multiple C–C and C–N bonds.

Table 4.8 Relative enthalpies of benzenoid heteroaromatic compounds

Compound	H^{o}_{rel} kJ/mol	Effect kJ/mol	Compound	H^{o}_{rel} kJ/mol	Effect kJ/mol
	76.3		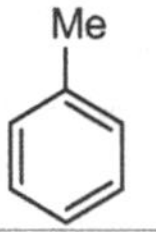 Me	64.0	−14.3
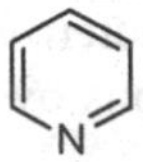 N	45.4	−30.9	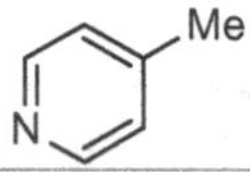 Me N	28.0	−17.4
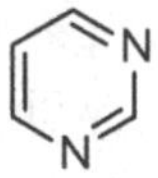 N N	13.6	−62.7	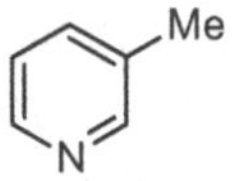 Me N	31.9	−13.5
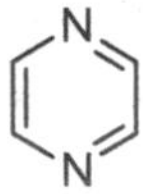 N N	13.1	−63.2	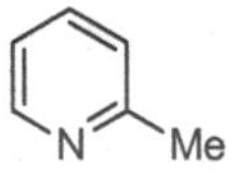 N Me	24.8	−20.4
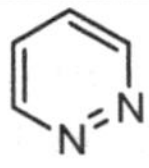 N=N	95.4	19.1	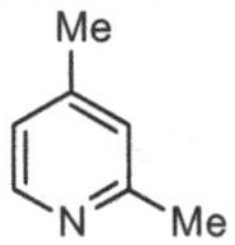 Me N Me	10.2	−35.2
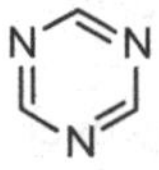 N N N	−45.3	−121.6	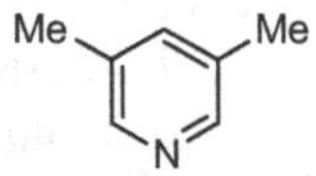 Me Me N	19.2	−26.2
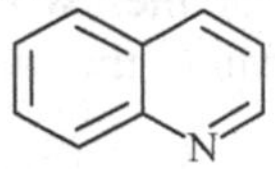 N	57.8	−38.7	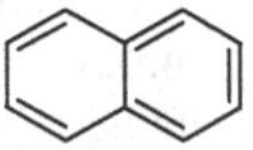	96.5	

CH_3-CH=CH_2	CH_3-CH=NH	HC≡CH	HC≡N	benzene	pyridine
82.3	−2.3	224.5	40.1	76.3	45.4

4.7 Nitro Compounds

Introduction of one or more nitro groups into organic compounds raises their energy. The H^o_{rel}-s listed in Table 4.9 clearly demonstrate it. The reason is the presence of the unfavorable N-O bonds in these compounds.

In the series of mononitro alkanes the H^o_{rel} of the first member nitromethane is the highest. Contribution of the nitro group to the energy of a compound is very high. For example, the nitro group in nitromethane increases the energy from −10.8 kJ/mol to ~126 kJ/mol. The difference is ~136 kJ/mol. This demonstrates that the N–O bonds of the nitro group are energetically very unfavorable.

The H^o_{rel}-s of nitroalkanes from nitro ethane to 1-nitro butane are almost the same as expected. The order of the alkyl groups, like in other kinds of compounds makes differences. Increasing order decreases the H^o_{rel}-s.

Et-NO_2	iso-Pr-NO_2	tBu-NO_2
124.1	105.7	89.5

Multiple nitro groups in the molecules further increase the energy as shown by the H^o_{rel}-s of nitromethanes.

CH_3-NO_2	O_2N-CH_2-NO_2	O_2N-$C(NO_2)_3$
125.7	286.5	708.1

The second nitro group in dinitromethane increases the energy relative to nitromethane by further ~160 kJ/mol. The H^o_{rel} of tetranitromethane is exceptionally high. The increment per nitro group is around 180 kJ/mol!

Table 4.9 Mononitro alkanes

Compound	H^o_{rel} kJ/mol	Compound	H^o_{rel} kJ/mol
Me–NO_2	125.7	iPr–NO_2	105.7
Et–NO_2	124.1	sec-Bu–NO_2	103.0
Pr–NO_2	121.3	tert-Bu–NO_2	89.5
Bu–NO_2	122.6		

Table 4.10 Nitro derivatives of benzene

Compound	H^o_{rel} kJ/mol	Compound	H^o_{rel} kJ/mol
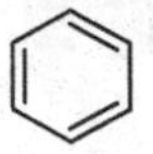	76.3	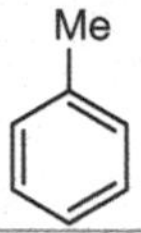 Me	64.0
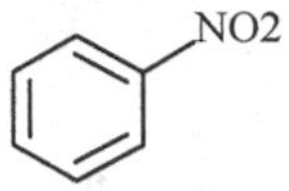 NO2	202.5	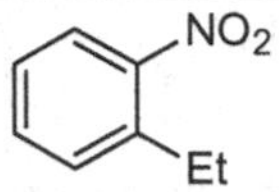 NO_2, Et	186.6
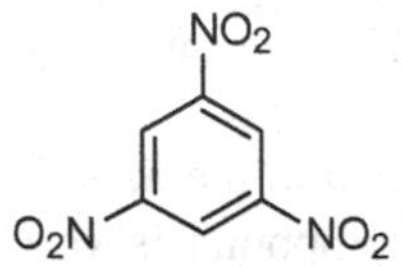 NO_2, O_2N, NO_2	471.5	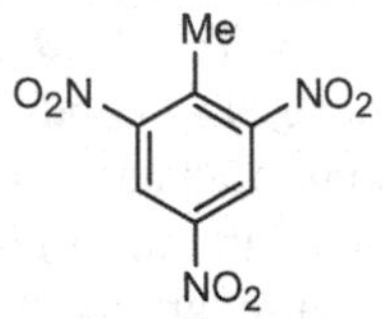 Me, O_2N, NO_2, NO_2	487.5

The energy of nitrobenzenes (Table 4.10) is also increased relative to benzene by the presence of nitro groups. The increment is about 120–130 kJ/mol per nitro-group. The H^o_{rel} of 2,4,6-trinitro-toluene, that is an explosive, is of course very high. The increment per nitro group is 141 kJ/mol.

4.8 Organic Nitrites and Nitrates

Organic nitrites are derivatives of the nitrous acid HNO_2. H^o_{rel}-s of a few alkyl nitrites are listed in Table 4.11. If their H^o_{rel}-s are compared to those of the corresponding nitroalkanes of Table 4.9 it can be observed, that their relative enthalpies are about the same.

The parent compound of alkyl-nitrates is nitric acid, HNO_3. The H^o_{rel}-s listed in Table 4.11 show that both the nitric acid itself and its alkyl derivatives are high energy compounds. The H^o_{rel}-s of the alkyl nitrates are even higher than those of the

Table 4.11 Relative enthalpies of organic nitrites and nitrates

Compound	H^o_{rel} kJ/mol	Compound	H^o_{rel} kJ/mol
Me–ONO	140.7	Me–ONO_2	210.1
Et–ONO	121.6	Et–ONO_2	197.2
iBu–ONO	103.0	Pr–ONO_2	197.7
		iPr–ONO_2	180.8
HNO_3	175.9	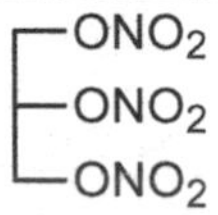 ONO_2, ONO_2, ONO_2	634.2

corresponding nitrites and nitro derivatives. The H^o_{rel} of the explosive glycerine trinitrate is particularly high. The energy increment per nitrate group is 211 kJ/mol, it is even higher than that per nitro group of tetranitromethane.

Reference

1. Furka Á (2009) Struct Chem 20:605–616

Chapter 5
Organosulfur Compounds

Sulfur differs from oxygen since it is a low energy element. Its H^o_{rel} is a considerable negative value that is in part the consequence of the fact that it refers to solid state [1].

O_2	S(solid)
251.7	−41.626

5.1 Thiols

While alcohols are derivatives of water, thiols can be deduced from hydrogen sulfide. When the H^o_{rel} of hydrogen sulfide is compared to that of water it can be seen that both compounds are low energy substances but the H^o_{rel} of water is much lower.

H-S-H	H-O-H
−18.4	−72.6

Since the H^o_{rel}-s of alcohols are higher than that of water, the H^o_{rel}-s of thiols are also expected to be higher than the H^o_{rel} of hydrogen sulfide. The comparisons below and the data of Table 5.1 fulfill this expectation.

H-S-H	Et-SH	Et-OH
−18.4	−3.0	−24.3

The H^o_{rel} of the first member of the series, as usual, differs from those of the rest of the members.

The next comparison demonstrates that the H^o_{rel} of thiols depends on order of the alkyl groups like in the case of alcohols.

Á. Furka, *The Structure Dependent Energy of Organic Compounds*, SpringerBriefs in Molecular Science, https://doi.org/10.1007/978-3-030-06004-6_5

Table 5.1 Relative enthalpies of n-alkyl thiols

Compound	H^o_{rel} kJ/mol	Compound	H^o_{rel} kJ/mol
Methanethiol	−0.6	1-Undecanethiol	−3.4
Ethanethiol	−3.0	l-Dodecanethiol	−3.5
1-Propanethiol	−4.2	1-Tridecanethiol	−3.4
1-Butanethiol	−3.8	1-Tetradecanehtiol	−3.4
1-Pentanethiol	−3.5	1-Pentadecanehtiol	−3.4
1-Hexanethiol	−3.5	l-Hexadecanehtiol	−3.4
1-Heptanethiol	−3.5	1-Heptadecanehtiol	−3.4
1-Octanethiol	−3.4	1-Octadecanehtiol	−3.4
1-Nonanethiol	−3.4	1-Nonadecanehtiol	−3.3
1-Decanethiol	−3.5	l-Eicosanethiol	−3.4

Et-SH −3.0   iPr-SH −12.5 $Me_2C(Et)$-SH −22.1

The H^o_{rel} of dithiols correspond to the sum of the H^o_{rel}-s of two thiols.

CH_3-CH_2-CH_2-CH_2-SH −3.8   HS-CH_2-CH_2-CH_2-CH_2-SH −7.1

The H^o_{rel}-s of cycloalkanethios closely correspond to those of the secondary alkanethiols. This is revealed by comparing the H^o_{rel}-s of cyclopentanethiol and cyclohexanethiol to those of cyclopentane, cyclohexane and 2-propanethiol.

cyclopentane 25.9   cyclopentane-SH 13.6   iPr-SH −12.5   cyclohexane 0.6   cyclohexane-SH −13.8

Substitution of one hydrogen atom of benzene by SH group reduces the energy by 13 kJ/mol. The same substitution by hydroxyl group, however, reduces the H^o_{rel} by a much higher value: ~53 kJ/mol. This allows to conclude that the effect of the SH group on energy of thiophenols is similar to that of hydroxyl group in phenols but it is much weaker. The weaker effect is in part the consequence of the lower energy reducing property of the S–H bond compared to the O–H bond as reflected by comparing the H^o_{rel}-s of thiols and alcohols.

Benzene 76.3 | Thiophenol (SH) 63.2 | Phenol (OH) 22.8

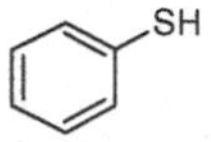

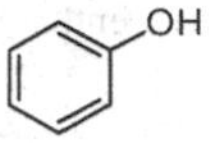

Et-SH −3.0 | Et-OH −24.3

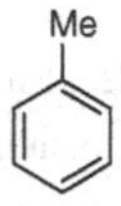

Toluene (Me) 64.0 | Thiophenol (SH) 63.2

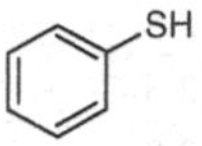

If the H^o_{rel}-s of toluene and tiophenol are compared it turns out that the energy reduction capability of the methyl and SH group is the same.

5.2 Organic Sulfides

The reference substances of the sulfur containing organic compounds are the non-branched R–CH_2–S–CH_2–R type dialkyl sulfides. Their H^o_{rel}-s are listed in Table 5.2. Being reference substances, the H^o_{rel}-s of dialkyl sulfides are very close to zero.

Table 5.3 demonstrates the H^o_{rel}-s of the methyl alkyl sulfides. As expected, the highest H^o_{rel} belongs to dimethyl sulfide. The H^o_{rel}-s of all the other methyl alkyl sulfides are lower positive values. The length of the second alkyl group replacing one of the methyl group has practically no effect on their H^o_{rel}-s.

Table 5.2 Relative enthalpies of dialkyl sulfides

Compound	H^o_{rel} kJ/mol	Compound	H^o_{rel} kJ/mol
Diethyl sulfide	0.8	Dodecyl propyl sulfide	0.3
Ethyl propyl sulfide	0.3	Propyl tridecyl sulfide	0.2
Butyl ethyl sulfide	−1.3	Propyl tetradecyl sulfide	0.3
Butyl propyl sulfide	0.2	Pentadecyl propyl sulfide	0.2
Ethyl pentyl sulfide	0.3	Hexadecyl propyl sulfide	0.3
Ethyl hexyl sulfide	0.4	Heptadecyl propyl sulfide	0.3
Ethyl heptyl sulfide	0.4	Dibutyl sulfide	−0.5
Ethyl octyl sulfide	0.3	Butyl pentyl sulfide	−0.5
Ethyl nonyl sulfide	0.4	Butyl hexyl sulfide	−0.5
Ethyl decyl sulfide	−1.2	Butyl heptyl sulfide	−0.6
Ethyl undecyl sulfide	0.5	Butyl octyl sulfide	−0.5
Ethyl dodecyl sulfide	0.4	Butyl nonyl sulfide	−0.5

Table 5.3 Relative enthalpies of methyl alkyl sulfides

Compound	H^o_{rel} kJ/mol	Compound	H^o_{rel} kJ/mol
Dimethyl sulfide	5.6	Dodecyl methyl sulfide	2.9
Methyl propyl sulfide	2.5	Methyl tridecyl sulfide	2.8
Methyl pentyl sulfide	3.7	Methyl tetradecyl sulfide	2.8
Hexyl methyl sulfide	2.7	Methyl pentadecyl sulfide	2.8
Methyl octyl sulfide	2.8	Hexadecyl methyl sulfide	2.8
Methyl nonyl sulfide	2.8	Heptadecyl methyl sulfide	2.8
Decyl methyl sulfide	2.8	Methyl octadecyl sulfide	2.9
Methyl undecyl sulfide	2.8	Methyl nonadecyl sulfide	2.9

Table 5.4 Higher order alkyl, allyl and benzyl sulfides

Compound	H^o_{rel} kJ/mol	Compound	H^o_{rel} kJ/mol
Ethyl isopropyl sulfide	−12.3	Di-tert-butyl sulfide	−40.3
Diisopropyl sulfide	−15.8	Allyl ethyl sulfide	79.5
Diisobutyl sulfide	−12.7	Allyl tert-butyl sulfide	56.3
Diisopentyl sulfide	−13.8	Benzyl methyl sulfide	72.5
tert-Butyl methyl sulfide	−16.1	Benzyl ethyl sulfide	65.5
tert-Butyl ethyl sulfide	−22.2	Dibenzyl sulfide	135.4

Higher order alkyl groups in sulfides considerably reduce the energy. This is demonstrated in Table 5.4. Increasing the order of the alkyl groups in one or both sides of the sulfur atom, from first to second and third order, gradually reduces the H^o_{rel} from zero to −40 kJ/mol.

Comparing the H^o_{rel} of allyl ethyl sulfide to that 1-hexene shows that the replacement of the C4 methylene group of 1-hexene by a sulfur atom only slightly reduces the value of H^o_{rel}.

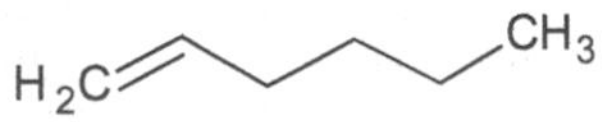

H_2C= ... CH_3 82.0 H_2C= ... S ... CH_3 79.5

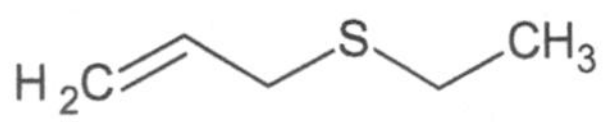

Below, the H^o_{rel} of ethyl phenyl sulfide is compared to those of benzene and ethyl phenyl ether. It can be seen that the H^o_{rel} of ethyl phenyl sulfide is lower than that of benzene but it is far to be as low as the H^o_{rel} of ethyl phenyl ether. The effect of the sulfur atom in aryl alkyl sulfides is similar to that of oxygen atom in aryl alkyl ethers: reduces the energy but to lower extent than does the oxygen atom.

76.3 70.0 58.9

The next comparison demonstrates that in diphenyl sulfide the sulfur atom shows about the same energy reducing effect as the methylene group in diphenylmethane.

2×76.3 = 152.6 133.7 128.7

5.3 Cyclic Sulfides

Table 5.5 demonstrates the H^o_{rel}-s of a series of cyclic sulfides from thiacyclopropane to thiacyclohexane. The strained ring representatives, thiacyclopropane and thiacyclobutane have high positive H^o_{rel}-s as expected.

These H^o_{rel}-s, however, are considerably lower than those of the corresponding cycloalkanes and oxacycloalkanes. This reflects that

sulfur better accommodates to rings strains than carbon, oxygen or nitrogen.

This is a special feature of sulfur that distinguishes it from both oxygen and nitrogen.

It has to be noted that the H^o_{rel} of thiacyclobutane in Table 5.5 was calculated from a heat of formation (18 kJ/mol) published in 2004. According to an older publication (from 1963), however, the heat of formation is 59.9 kJ/mol. The H^o_{rel} calculated from this value is 80.1 kJ/mol. This is also a lower value than the H^o_{rel} of cyclobutane and as a consequence also strengthens the special feature of sulfur mentioned above. In addition, it seems an even more realistic approach than the very low 38.2 kJ/mol in Table 5.5.

Before turning to the problem of stabilization energy of thiophene it seems worthwhile to mention the stabilization effect observable in thioacetals. As already

Table 5.5 Relative enthalpy of cyclic sulfides

Compound	H^o_{rel} kJ/mol	Compound	H^o_{rel} kJ/mol	Compound	H^o_{rel} kJ/mol	Compound	H^o_{rel} kJ/mol
	115.2	S	81.8	O	114.4	S	7.6
	109.2	S	38.2	O	107.2	S	−1.8

mentioned the structure dependent energies of acetals and ketals are lower than those of their parent oxo compounds. It is a question whether or not this behavior holds for thioacetals, too. As it is seen below, the H^o_{rel} of bis(ethylthio)methane is also lower than that of formaldehyde but not as low as the H^o_{rel} of diethoxymethane. The energy reducing capability of sulfur in these compounds is not as effective as that of oxygen.

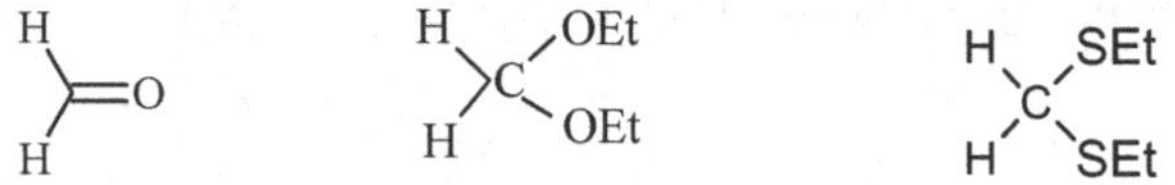

30.6 −14.9 −1.9

5.3.1 *Stabilization Energy of Thiophene*

Thiophene is considered to be an aromatic compound. The first step in deducing the stabilization energy is comparing the H^o_{rel}-s of thiophene and cyclopentadiene.

70.4^{1963} (172.4^{1991}) 150.1

The problem is, like in the case of pyrrole, that there are available two considerably different experimental heats of formation. One of them was published in 1963 (116.4 kJ/mol) and the other in 1991 (218.4 kJ/mol). Of course the H^o_{rel}-s deduced from them also differ. If the older and lower value of the two H^o_{rel}-s is used in the comparison with the H^o_{rel} of cyclopentadiene, 79.7 kJ/mol comes out as energy reduction in thiophene. If the newer and higher value is compared, instead of reduction a ~22 kJ/mol increase comes out that is absolutely unacceptable.

For this reason the estimation is continued with the 79.7 kJ/mol energy reduction that comes out using the lower H^o_{rel} value of thiophene. This energy reduction is supposed to be the sum of two values: the stabilization energy and the reduction coming from replacement by sulfur of the methylene group of cyclopentadiene. This later value can be deduced by comparing the H^o_{rel}-s of 1,4-pentadiene and divinyl sulfide.

H_2C=CH–CH$_2$–CH=CH_2 165.0 H_2C=CH–S–CH=CH_2 103.4

The comparison reveals that the replacement brings about a 61.6 kJ/mol reduction. In order to get the stabilization energy this value needs to be subtracted from the 79,7 kJ/mol.

$$\text{Stabilization energy} = 79.7\text{–}61.6 = 18.1\ \text{kJ/mol}$$

This value looks very low for stabilization energy. This may be the consequence of the uncertain heat of formation of thiophene. There is another factor that may add to the problem. The H^o_{rel} of divinyl sulfide also looks uncertain considering the comparisons below.

128.7 133.7

H_2C CH_2 165.0 H_2C S CH_2 103.4

Replacing the methylene group of diphenylmethane by sulfur an energy increase of 5 kJ/mol comes out. Replacing the central methylene group of 1,4-pentadiene by sulfur the result is a very high, 61.6 kJ/mol energy reduction. The two results are incompatible and questions the acceptability of the H^o_{rel} of divinyl sulfide.

Considering the problems with the H^o_{rel}-s of both thiophene and divinyl sulfide one has to conclude that the available data are not accurate enough to make possible the reliable estimation of stabilization energy.

5.4 Thioaldehydes

As the H^o_{rel}-s of thioformaldehyde and thioacetaldehyde demonstrate below the thioaldehydes are high energy compounds. In order to see the energy determining property of the carbon-sulfur double bond the H^o_{rel}-s of thioaldehides are compared to those of aldehydes and alkenes.

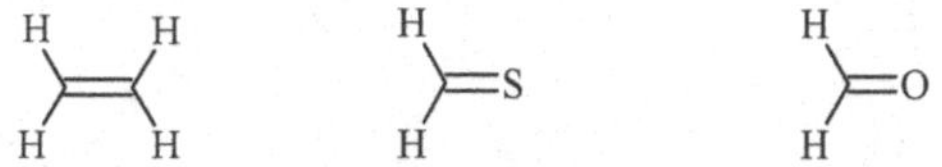

93.5 97 (69) 30.6

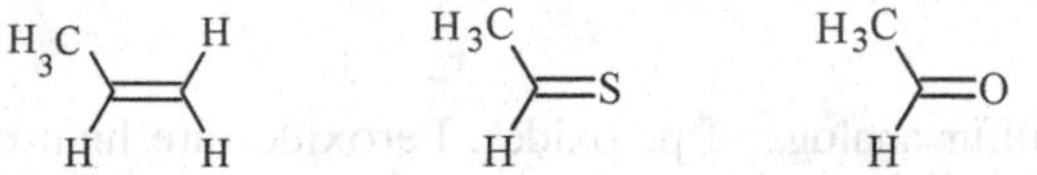

H_3C 82.3 H_3C 50 H_3C 0.7

If the carbon-sulfur double bond is compared to the carbon oxygen double bond the former bond looks highly unfavorable. When the C=S and the C=C bond is compared it has to be taken into account that two different heats of formation were published for thioformaldehyde: one in 1982 (90 kJ/mol) and a newer one in 1993

(118 kJ/mol). The H^o_{rel} calculated from the newer ΔH^o_f is even higher than that of ethylene that looks unlikely. The H^o_{rel} coming from the older ΔH^o_f (above in parenthesis) is between the H^o_{rel}-s of formaldehyde and ethylene, closer to ethylene. Comparing the H^o_{rel}-s of thioacetaldehide, acetaldehyde and propene, supports the later result. Considering the energy of the three double bonds the final conclusion is:

$$C = C > C = S > C = O$$

5.5 Carbothioic S-Acids and Esters

In carbothioic S-acids the hydroxyl oxygen of the carboxylic acids is replaced by sulfur. As exemplified by comparing below the H^o_{rel}-s of acetic acid and thioacetic acid the O to S replacement causes dramatic change in H^o_{rel}: it is increased from −139.1 to −56.5 kJ/mol. The reason is that the interaction of the SH group with the carbonyl group is less efficient in reducing the energy than that of the OH group.

$H_3C-C(=O)OH$ −139.1 $H_3C-C(=O)SH$ −56.5

Comparisons of the pairs of esters below reveals that this property of the sulfur atom shows up in the H^o_{rel}-s of the esters of the carbothioic S-acids, too.

$H-C(=O)O-CH_2CH_3$ −48.1 $H-C(=O)S-CH_2CH_3$ −31.6 $H_3C-C(=O)O-CH_2CH_3$ −110.7 $H_3C-C(=O)S-CH_2CH_3$ −61.1

5.6 Disulfides

Disulfides are sulfur analogs of peroxides. Peroxides are high energy compounds and the sulfides as the examples below show, are rather low energy substances. While the H^o_{rel} of dimethyl peroxide is a high positive value, the relative enthalpy of dimethyl disulfide is expressed by a negative number. The low energy of the dialkyl disulfides is further supported by data of Table 5.6.

Table 5.6 Relative enthalpies of dialkyl disulfides

Compound	H^o_{rel} kJ/mol	Compound	H^o_{rel} kJ/mol
Dimethyl disulfide	−22.7	Diheptyl disulfide	−33.2
Diethyl disulfide	−31.9	Dioctyl disulfide	−33.2
Dipropyl disulfide	−33.3	Dinonyl disulfide	−33.2
Dibutyl disulfide	−33.3	Didecyl disulfide	−33.1
Dipentyl disulfide	−33.2	Diisobutyl disulfide	−45.2
Dihexyl disulfide	−33.2	Di-tert-butyl disulfide	−72.0

H_3C–O–O–CH_3 210.9 H_3C–S–S–CH_3 −22.7

As Table 5.6 demonstrates the H^o_{rel}-s of n-dialkyl disulfides are close to −33 kJ/mol. The first member of the series, dimethyl disulfide is again an exception, its H^o_{rel} is higher by ~10 kJ/mol than those of the rest of the n-dialkyl disulfides listed in the Table 5.6.

The H^o_{rel}-s of the last two compounds in the table, diisobutyl disulfide and di-tert-butyl disulfide reflects that branching in the alkyl groups and particularly the order of the alkyl groups have a considerable energy reducing effect.

In comparison to peroxides, the low energy of dialkyl disulfides show that considering the energy, the S-S bond is not unfavorable and this sharply distinguishes sulfur from oxygen.

5.7 Sulfoxides, Sulfones, Sulfites and Sulfates

Sulfoxides and sulfones are oxidized sulfides. The H^o_{rel}-s of sulfoxides and sulfones are listed in Table 5.7.

Et-S-Et diethyl sulfide Et-SO-Et diethyl sulfoxide Et-SO_2-Et diethyl sulfone

The comparisons below demonstrate that mono-oxidation does not change significantly the H^o_{rel}-s of sulfides. The effect on energy of the SO group in sulfoxides can't be compared to that of the CO group in ketones that considerably reduces the energy.

Table 5.7 Relative enthalpies of sulfoxides, sulfones, sulfites and sulfates

Sulfoxide	H^o_{rel} kJ/mol	Sulfone	H^o_{rel} kJ/mol
Me–SO–Me	24.1	Me–SO_2–Me	−78.0
Et–SO–Et	4.7	Et–SO_2–Me	−94.2
Pr–SO–Pr	−3.4	Et–SO_2–Et	−94.5
tBu–SO–Et	−22.7	Pr–SO_2–Pr	−96.8
Et–SO–CH_2–CH=CH_2	84.0	iPr–SO_2–Me	−96.2
Ph–SO–Ph	134.2	Ph–SO_2–Ph	36.7
Sulfite	H^o_{rel} kJ/mol	Sulfate	H^o_{rel} kJ/mol
Me_2SO_3	−62.7	H_2SO_4	−308.4
$EtMeSO_3$	−82.6	Me_2SO_4	−140.1
Et_2SO_3	−90.1	Et_2SO_4	−168.3
Pr_2SO_3	−84.8	Pr_2SO_4	−162.7
		Bu_2SO_4	−158.2

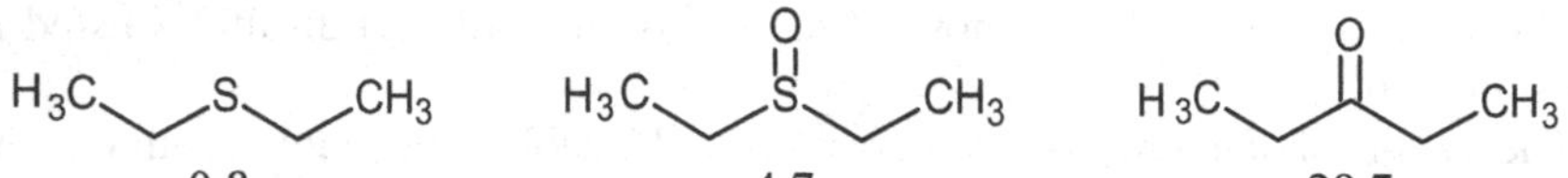

0.8 4.7 −29.7

The energy of sulfones, however, differs from both sulfides and sulfoxides. The following comparison shows that sulfones are low energy compounds. The presence of the second oxygen atom in the molecule reduces the energy by ~100 kJ/mol!

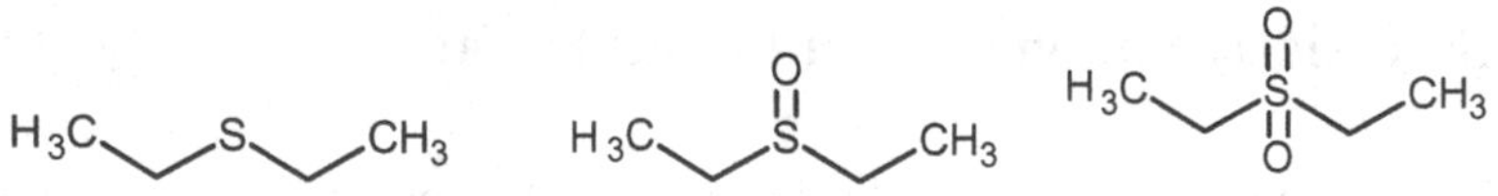

0.8 4.7 −94.5

The H^o_{rel} of allyl ethyl sulfoxide allows to conclude that replacement of one of the ethyl groups in diethyl sulfoxide by an allyl group increases the H^o_{rel} as expected, by a value close to the H^o_{rel} of propene.

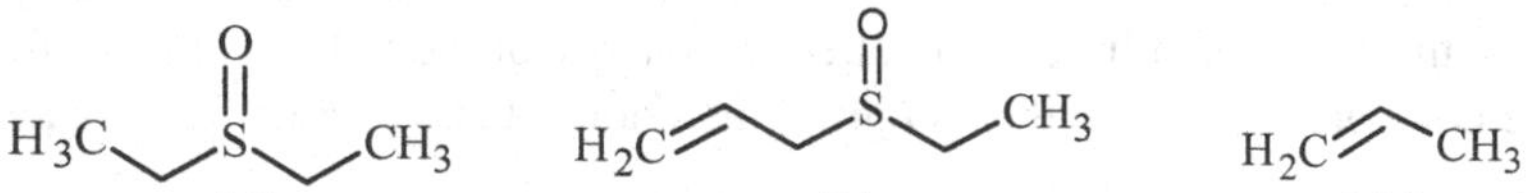

4.7 84 82.3

The H^o_{rel} of diphenyl sulfoxide is nearly equal to the sum of the H^o_{rel}-s of diethyl sulfoxide and diphenylmethane (133.4 kJ/mol).

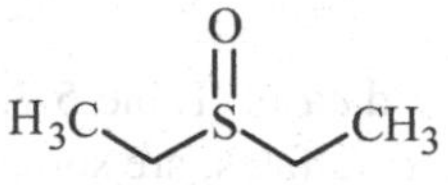

4.7 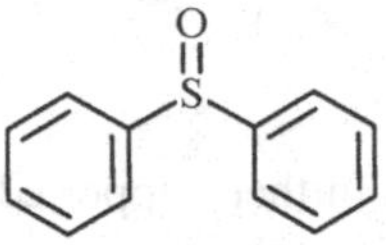 134.2 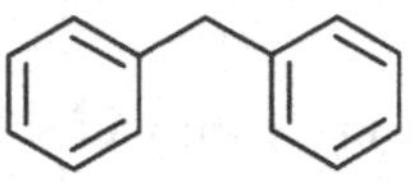 128.7

The H^o_{rel}-s of dialkyl sulfones are high negative values as demonstrated by the data of Table 5.7. The H^o_{rel} of the first member, dimethyl sulfone, is again an exception.

The H^o_{rel} of diphenyl sulfone also nearly equals the sum of the H^o_{rel} of diethyl sulfone plus the H^o_{rel}-s of diphenylmethane (34.2 kJ/mol) demonstrating that the energy of diphenyl sulfone is entirely determined by its three functional groups: the SO_2 group and the two benzene rings.

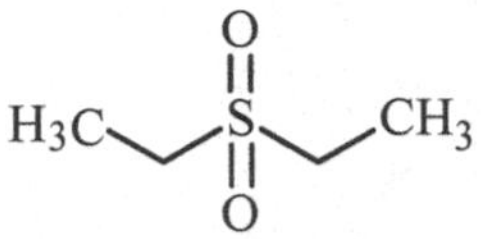

–94.5 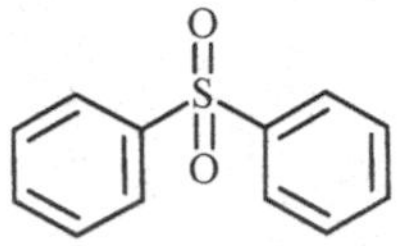 36.7 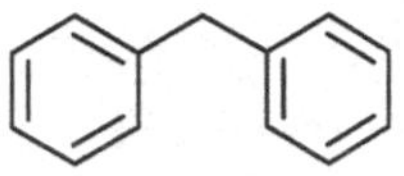 128.7

The alkyl sulfites and sulfates are alkyl derivatives of sulfurous acid and sulfuric acid, respectively.

R – O – SO – OR	RO – SO_2 – OR
diethyl sulfite	diethyl sulfate

As reflected by the H^o_{rel}-s of Table 5.7 dialkyl sulfites and dialkyl sulfates are low energy compounds.

As the comparison below demonstrates, the H^o_{rel}-s of sulfites are close to those of sulfones while the H^o_{rel}-s of sulfates are much lower. The sulfates are esters of sulfuric acid and the low energy of diethyl sulfate, for example, is the consequence of the exceptionally low energy of the sulfuric acid (−308.4 kJ/mol). The data also show that esterification of sulfuric acid highly increases the energy.

–94.5 –90.1 –168.3

The low energy of sulfates is further supported by the data of Table 5.7. The H^{o}_{rel}-s of the dimethyl derivatives in both groups, sulfites and sulfates, are somewhat higher than those of the rest of the members of the two groups.

Reference

1. Furka Á (2009) Struct Chem 20:605–616

Chapter 6
Organohalides

6.1 Halogens

As reflected by their relative enthalpies, the four halogens occurring in organic compounds have different energies in their elemental form. As Table 6.1 shows the highest energy belongs to fluorine, then going in direction chlorine, bromine and iodine the energy gradually decreases. In fact, among all the elements occurring in organic compounds fluorine has the highest energy. This is in accordance with its properties observed by experience. Concerning the energy, on the other end of the series of elements mentioned in this book, nitrogen (N_2) is found. Its relative enthalpy is −178.6 kJ/mol [1].

The hydrogen halides are low energy compounds. Their H^o_{rel}-s are also gradually changing. The lowest H^o_{rel} belongs to hydrogen fluoride the highest to hydrogen iodide but its H^o_{rel} is still a negative value.

6.2 Alkyl Halides

It has long been observed that in the stepwise fluorination of methane the heats of reactions are increasing step by step. This allows concluding that in these reactions gradually lower and lower energy compounds are formed. The same trend is clearly observed by directly comparing to each other the H^o_{rel}-s of the fluorinated derivatives of methane listed in Table 6.2. This reflects the special feature of fluorine, since in contrast to this, in the case of the other three halogens the H^o_{rel}-s of the compounds are gradually increasing with the degree of substitution. This is clearly reflected by the rest of the H^o_{rel}-s of Table 6.2.

Á. Furka, *The Structure Dependent Energy of Organic Compounds*, SpringerBriefs in Molecular Science, https://doi.org/10.1007/978-3-030-06004-6_6

Table 6.1 Relative enthalpies of halogens and hydrogen halides

Element	State	H^o_{rel} kJ/mol	Compound	H^o_{rel} kJ/mol
F_2	G	396.12	HF	−51.3
Cl_2	G	94.22	HCl	−23.5
Br_2	L	8.16	HBr	−10.4
I_2	S	−107.6	HI	−5.7

Table 6.2 Relative enthalpies of halogen derivatives of methane

Compound	H^o_{rel} kJ/mol	Change kJ/mol	Compound	H^o_{rel} kJ/mol	Change kJ/mol
CH_3F	1.5	12.3	CH_3Br	8.7	19.5
CH_2F_2	−36.2	−34.7	CH_2Br_2	24.6	15.9
CHF_3	−104.4	−68.2	$CHBr_3$	36.2	11.6
CF_4	−163.6	−59.2	CBr_4	43.7	7.5
CH_3Cl	3.2	14	CH_3I	2.5	13.3
CH_2Cl_2	19.4	16.2	CH_2I_2	31	28.5
$CHCl_3$	38.9	19.5	CHI_3	48.4	17.4
CCl_4	65.2	26.3			

It is worthwhile to remind that the gradual decrease of the energy of the compound formed in stepwise substitution the hydrogen atoms of the same carbon atom by fluorine was observed with oxygen, too. The energy reductions in the steps of alkoxy substitutions are lower than those in fluorine ones.

	CH_3F		CH_2F_2		CHF_3		CF_4
	1.5		−36.2		−104.4		−163.6
Differences:		37.7		68.2		59.2	

	H_3C–OMe		$H_2C(OMe)_2$		H–C(OMe)$_3$		MeO–C(OMe)$_3$
	26.4		8.6		−28.3		−77.3
Differences:		17.8		36.9		49	

	H_3C–OEt		$H_2C(OEt)_2$		H–C(OEt)$_3$		EtO–C(OEt)$_3$
	14.8		−14.9		−65.3		−129.3
Differences:		29.7		50.4		64	

As exemplified by the H^o_{rel}-s below, in addition to the fluorine derivatives of methane, the high stabilization by stepwise fluorine substitution is also observed with other compounds containing multi-fluorinated carbon atoms.

CH_3--CH_2F	CH_3--CHF_2	CH_3--CF_3
−0.5	−56.3	−131.9

The above data allow to conclude, that

it is a remarkable property of fluorine and oxygen: the energy is reduced with the number of single bonds they make with the same carbon atom. A similar but a much weaker trend is observed with nitrogen.

Other data in Table 6.2 demonstrates that if one fluorine atom in CF_4 is replaced by one chlorine atom, one bromine atom or one iodine atom, the energy of the compound increases by about the same value: ~90 kJ/mol. This is reflected by nearly the same H^o_{rel}-s of the three compounds.

CF_4	CF_3Cl	CF_3Br	CF_3I
−163.6	−76	−73.5	−72.4

It also seems worthwhile to compare the H^o_{rel} of tetrafluoromethane to that of carbon dioxide. It can be seen that the two values are almost the same.

CF_4	O=C=O
−163.6	−164.6

Table 6.3 shows the effects of halogen substitutions on primary, secondary and tertiary carbon atoms on the H^o_{rel}-s of alkyl halides, that can be compared to those of methyl substitutions.

Substitution on primary carbon atom of n-alkanes by either methyl group or any halogen atom practically (by definition) does not change the value of H^o_{rel} since it leads to a reference compound. If the substitution occurs on secondary carbon, the energy reducing effect of halogens, except chlorine and iodine, only slightly

Table 6.3 Relative enthalpies of secondary and tertiary alkyl halides

X	Me–CH_2–X H^o_{rel} kJ/mol	Me_2CH–X H^o_{rel} kJ/mol	Me_3C–X H^o_{rel} kJ/mol
Me	1.5	−8.6	−19.5
F	−0.5	−7.1	−40.8
Cl	−1.6	−15.7	−32.0
Br	3.0	−9.4	−25.6
I	−0.8	−11.8	−23.2

exceeds that of the methyl group. In the case of substitution on tertiary carbon, however, the stabilization effects of all halogen substitutions are stronger than that of methyl substitution and are increasing from iodine to fluorine.

Examples for the relative enthalpies of cycloalkyl halides are shown below. The energy reducing effect in fluorocyclohexane seems to be considerably stronger than that showing up at substitution on a secondary carbon atom of an open chain alkane in Table 6.3 (7.1 kJ/mol).

F Cl Br Br

0.6 –36.6 –17.4 –21.5

6.3 Alkenyl Halides

The data listed in Table 6.4 can be used to demonstrate how the halogen substituents of ethylene modify the energy. The stabilization and destabilization effects can be compared to those of the methyl groups.

The H^{o}_{rel} of vinyl fluoride is close to that of propene showing that the stabilizing effect of fluorine is practically the same as that of the methyl group. Chlorine and bromine in vinyl chloride and vinyl bromide brings about destabilization. A double substitution by fluorine on the same carbon atom of ethylene causes stabilization that exceeds that of two methyl groups. A third fluorine atom reduces the magnitude

Table 6.4 Relative enthalpies of alkenyl halides

Compound	H^{o}_{rel} kJ/mol	Effect kJ/mol	Compound	H^{o}_{rel} kJ/mol	Effect kJ/mol
$CH_2=CH_2$	93.5		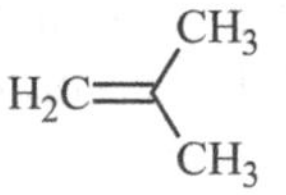 $H_2C=C(CH_3)_2$	64.6	−28.9
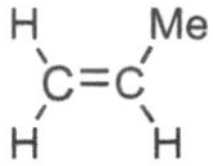 $H_2C=CH(Me)$	82.3	−11.2	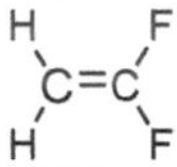 $H_2C=CF_2$	48.7	−44.8
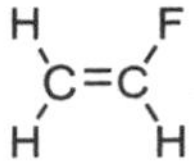 $H_2C=CHF$	81.6	−11.9	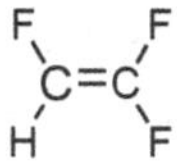 $HFC=CF_2$	74.4	−19.1
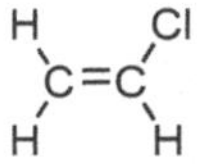 $H_2C=CHCl$	101.7	8.2	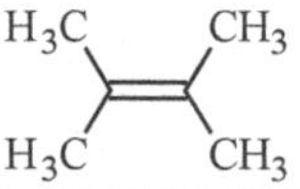 $(H_3C)_2C=C(CH_3)_2$	53.4	−40.1
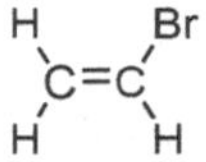 $H_2C=CHBr$	102.0	8.5	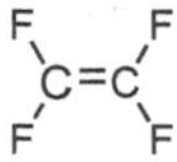 $F_2C=CF_2$	88.0	−5.5

Table 6.5 Chloro-derivatives of ethylene

Compound	H^o_{rel} kJ/mol	Effect of substitution kJ/mol	Compound	H^o_{rel} kJ/mol	Effect of substitution kJ/mol
H H H H	93.5		H CH_3 H H	82.3	−11.2
H Cl H H	101.7	8.2	Cl Cl Cl Cl	128.6	35.1
H Cl H Cl	93.3	−0.2	Cl Cl H H	93.9	0.4
Cl Cl H Cl	111.5	18.0	Cl H H Cl	96.2	2.7

of stabilization. Although four fluorine substituents also cause some stabilization but this is much lower than that of four methyl groups.

The H^o_{rel}-s of chloro derivatives of ethylene are listed in Table 6.5. As shown, the mono substitution results in an energy increase that looks more considerable if compared to the stabilizing substitution by methyl group. Substitution by chlorine of two hydrogens on the same carbon atom, however, leads to no change in H^o_{rel} while the two methyl groups in 1,1-dimethyl substituted ethylene brings about a considerable reduction in H^o_{rel} (28.9 kJ/mol in Table 6.4).

Substitution of three hydrogen atoms in the molecule by chlorine causes destabilization by 18 kJ/mol. Substitution of all the four hydrogen atoms brings about a moderately high energy increase (35.1 kJ/mol). The tetrachloro substitution, however, looks even more disadvantageous if the energy increase is compared to the stabilization effect of the four methyl groups in the tetramethylethylene (−40.1 kJ/mol in Table 6.4).

It seems worthwhile to compare the H^o_{rel}-s of the following five perfluoro compounds: tetrafluoromethane, hexafluoroethane, tetrafluoroethylene, perfluoro-2-methylpent-2-ene and dodecafluorocyclohexane.

The H^o_{rel}-s of the compounds are found right under the structures. In the second line are the total stabilization energies in kJ/mol relative to the parent hydrocarbons and the in the third lines stabilization energies per the number of fluorine atoms. The highest stabilization energy per fluorine atoms belongs to tetrafluoromethane, the second to hexafluoroethane and the lowest to tetrafluoroethylene.

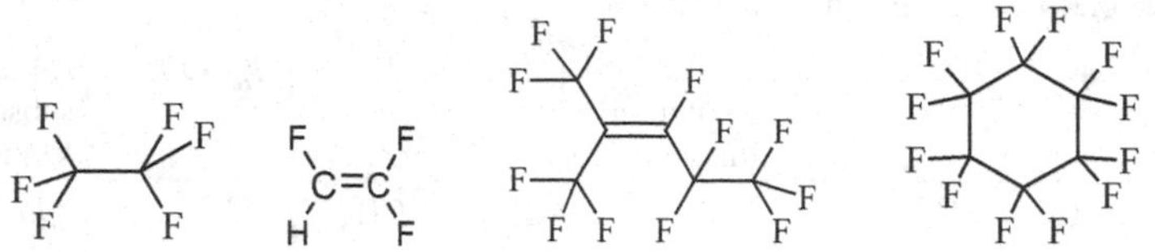

CF4				
−163.6	−184.9	88.0	−259.3	−231.3
174.4	184.9	5.5	320.9	231.9
43.6	30.8	1.4	26.7	19.3

The hexafluorobenzene (Table 6.6) also shows low stabilization energy per number of fluorine atoms: 6.8 kJ/mol. It is interesting that the H^o_{rel} of perfluoro-2-methylpent-2-ene that contains a double bond is lower than that of dodecafluorocyclohexane that is a saturated compound.

6.4 Aryl Halides

The relative enthalpies of a few halogen derivatives of benzene are listed in Table 6.6.

Table 6.6 Relative enthalpies of halogen derivatives of benzene

Compound	H^o_{rel} kJ/mol	Effect of substitution kJ/mol	Compound	H^o_{rel} kJ/mol	Effect of substitution kJ/mol
	76.3		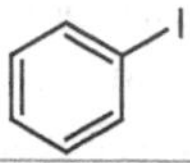 I	80.3	4.0
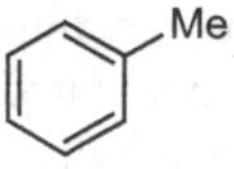 Me	64.0	−12.3	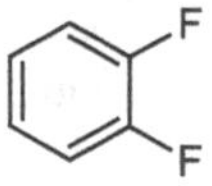 F, F	51.5	−24.8
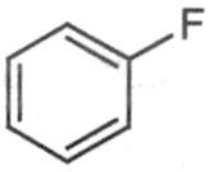 F	53.1	−23.2	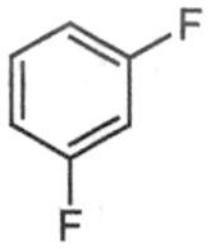 F, F	36.0	−40.3
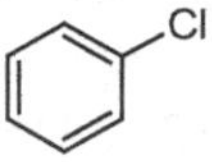 Cl	70.6	−5.7	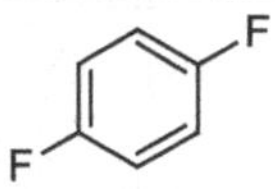 F, F	38.8	−37.5
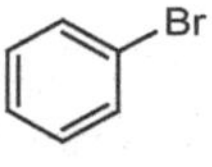 Br	80.7	4.4	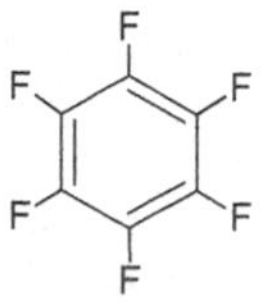 F, F, F, F, F, F	35.4	−40.9

Table 6.7 Chlorobenzenes

Compound	H^o_{rel} kJ/mol	Effect of substitution kJ/mol	Compound	H^o_{rel} kJ/mol	Effect of substitution kJ/mol
	76.3		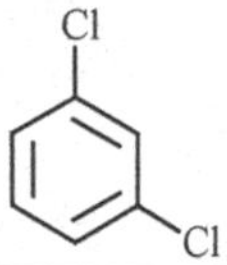 Cl, Cl	70.5	−5.8
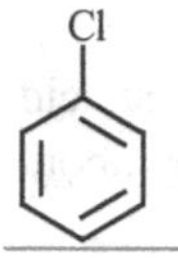 Cl	70.6	−5.7	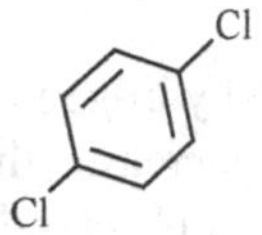 Cl, Cl	67.1	−9.2
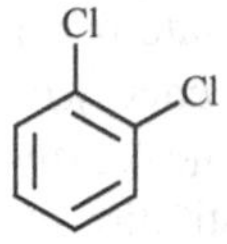 Cl, Cl	74.1	−2.2	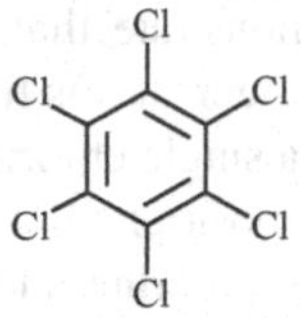 Cl, Cl, Cl, Cl, Cl, Cl	111.8	35.5

The effect of the halogen atoms on energy can be compared to that of a methyl group. The methyl group in toluene reduces the energy by ~12 kJ/mol. The energy reducing effect of the fluorine atom in fluorobenzene is higher: ~23 kJ/mol. The stabilization effect of the chlorine atom in chlorobenzene is only 5.7 kJ/mol, less than that of the methyl group. Both the bromine atom in bromobenzene and the iodine atom in iodobenzene increases H^o_{rel} by about ~4 kJ/mol. The second fluorine atom in difluorobenzenes further reduces the energy. The strongest effect (~37 and −40 kJ/mol) is observed when the two fluorine atoms occupy 1,3 or 1,4 position. The stabilization per fluorine atom is 18–20 kJ/mol/fluorine atom. The six fluorine atoms in hexafluorobenzene bring about a total stabilization of only 40.9 kJ/mol (6.8 kJ/mol/number of fluorine atoms).

Table 6.7 shows the H^o_{rel}-s of dichlorobenzenes. The values of the stabilizing effects are close to that of chlorobenzene except 1,4–dichlorobenzene, that shows somewhat higher stabilization. The H^o_{rel} of hexachlorobenzene, however, reflects the destabilizing effect of the six chlorine atoms (5.9 kJ/mol/number of chlorine atoms).

6.5 Halogen Derivatives of Carboxylic Acids and Esters

Effect of chlorine substituents on the energy of chloroacetic acids is demonstrated by data of Table 6.8. The H^o_{rel} of monochloroacetic acid shows that the chlorine substituent increases the energy by ~22 kJ/mol. The additional two chlorine atoms in trichloroacetic acid further increase the energy by ~53 kJ/mol. One can

Table 6.8 Chloroacetic acids

Compound	H^o_{rel} kJ/mol	Effect of substitution kJ/mol	Compound	H^o_{rel} kJ/mol
CH_3–COOH	−139.1		Propyl chloroacetate	−87.2
$ClCH_2$–COOH	−117.3	21.8	Butyl dichloroacetate	−71.7
Cl_3C–COOH	−64.6	74.5	Propyl trichloroacetate	−29.3

conclude that substitution of the hydrogen atoms in the methyl group of acetic acid by chlorine atoms considerable increases the energy (each chlorine atom by about 25 kJ/mol). A similar tendency is observed at chloroacetic acid esters.

The data of Table 6.9 demonstrate that substitution of the hydrogen atoms of benzoic acid in the ring by halogen atoms has an observable effect on the energy of the molecule. Substitution by a single chlorine atom reduces the energy. The energy reduction is most effective in positions 3 and 4 (~24–26 kJ/mol). Substitution in position 4 by iodine, bromine, chlorine and fluorine, the energy reducing effect is gradually increasing from ~13 to ~28 kJ/mol. Substitution of all of the five hydrogens in the ring by fluorine, significantly increases the energy.

6.6 Carboxylic Acid Halides

Carboxylic acid halides are reactive compounds. The high reactivity is not reflected by their H^o_{rel}-s. They are in fact low energy compounds that can be attributed to the interaction of the halogen atom and the carbonyl group. This is made clear by comparing the H^o_{rel}-s of the following compounds:

H_3C–CO–CH_3	H_3C–CO–F	H_3C–CO–Cl	H_3C–CO–Br	H_3C–CO–I
−29.9	−101.8	−51.8	−41.2	−34.9

Acetone is a low energy compound but replacing one of the methyl groups by a halogen atom, the value of the H^o_{rel} drops considerably. The degree of stabilization depends on the halogen: fluorine 71.8, chlorine 21.9, bromine 11.3 and iodine 5 kJ/mol. The energy reducing effect of the chlorine atom in the acid halides is about one third of that of the fluorine and a further, but gradually decreasing reduction is observed toward bromides and iodides. Thus the relative enthalpy of acetyl iodide is close to that of acetone.

Table 6.9 Halogen derivatives of benzoic acids

Compound	H^o_{rel} kJ/mol	Effect kJ/mol	Compound	H^o_{rel} kJ/mol	Effect kJ/mol
Benzoic acid	−68.0		4-Fluorobenzoic acid	−95.9	−27.9
2-Chlorobenzoic acid	−71.6	−3.6	4-Bromobenzoic acid	−85.8	−17.8
3-Chlorobenzoic acid	−93.8	−25.8	4-Iodobenzoic acid	−81.3	−13.3
4-Chlorobenzoic acid	−92.1	−24.1	Pentafluorobenzoic acid	−44.5	23.5

It is interesting to note that the H^o_{rel} of acetyl fluoride is close to that of ethyl acetate. This demonstrates that the strength of the energy reducing interaction of the fluorine and oxygen atoms with the carbonyl group in these compounds is nearly the same. This puts acetyl fluoride close to the low energy range of esters.

H_3C–C(=O)F: −101.8 H_3C–C(=O)–O–CH_2CH_3: −110.7

The H^o_{rel}-s of benzoyl halides are compared to that of acetophenone. In the case of benzoyl chloride a weak stabilization (17.5 kJ/mol) shows up, while the H^o_{rel}-s of benzoyl bromide and benzoyl iodide are practically the same as that of acetophenone.

C_6H_5–C(=O)CH_3: 30.1 C_6H_5–C(=O)Cl: 12.6 C_6H_5–C(=O)Br: 30.2 C_6H_5–C(=O)I: 31.4

The following comparisons demonstrate that substituting step by step the hydrogen atoms in acetyl chloride by chlorine atoms, gradually increases the H^o_{rel}-s. A similar trend was observed with chloroacetic acids, too.

CH_3–C(=O)Cl: -51.8 $ClCH_2$–C(=O)Cl: -27.7 Cl_2CH–C(=O)Cl: 1.8 Cl_3C–C(=O)Cl: 31.8

Another structural factor that has an unfavorable effect on the energy of carboxylic acid chlorides is the direct bonding of the two carbonyl groups in oxalyl dichloride. Although there are possibilities for two chlorine-carbonyl interactions in the molecule, like in two molecules of acetyl chloride, its H^o_{rel} is only −35.5 kJ/mol, much higher than the sum of the H^o_{rel}-s of two molecules of acetyl chlorides. The reason is the vicinal position of the two carbonyl groups.

CH_3–CO–Cl

CH_3–CO–Cl

2 x –51.8 = –103.6

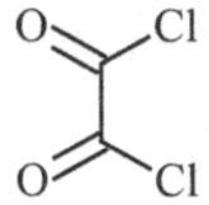

–35.5

6.7 Carbonyl Halides

In carbonyl halides two halogen atoms are bonded to the carbonyl group. Since there are possibilities for two halogen-carbonyl interactions in the molecules, lower H^o_{rel}-s are expected than in the case of the corresponding carboxylic acid halides.

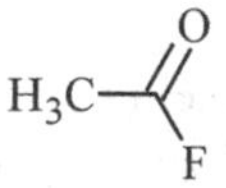

–101.8 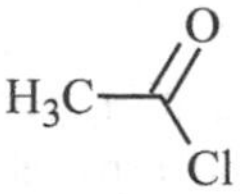 –51.9 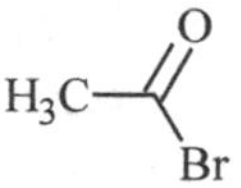 –41.4

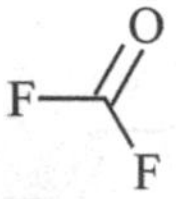

–140.8 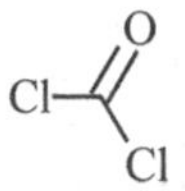 –21.6 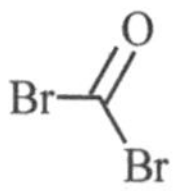 –2.3

The data above show that this expectation is fulfilled only in the case of carbonyl fluoride. The H^o_{rel}-s of carbonyl chloride and carbonyl bromide are not lower but considerably higher than the H^o_{rel}-s of the corresponding acetyl halides.

Reference

1. Furka Árpád (2009) Struct Chem 20:605–616

Chapter 7
Radicals, Cations, and Anions

Like in the case of compounds, the relative enthalpies of radicals, cations and anions calculated from their heats of formation [1, 2] can be used to characterize their energy. The H^o_{rel}-s do not depend on composition and, as a consequence can be compared to each other or to the H^o_{rel}-s of their parent compounds without any restriction. As mentioned, the H^o_{rel}-s of the parent compounds of radicals, cations and anions show only the effect of structures on energy. This is not completely true for radicals, cations and anions since their relative enthalpy, in addition to the effects on energy of the structural varieties, also depend on the energy needed for dissociation. For better comparisons, this strong effect needs to be removed. This is accomplished by subtracting the H^o_{rel} of methyl radical, methyl cation and methyl anion from the H^o_{rel}-s of radicals, cations and anions, respectively. The methyl radical, cation and anion in this respect can be regarded as structureless entities. They have no functional groups that could have influence on energy. The quantity that reflects the effect of structure on the energy of radicals, cations or anions is the remainder of the above subtractions. This case it is called Structure Dependent Energy and is abbreviated by SDE [3].

The SDE-s can answer the question whether the same structural elements in radicals, carbocations or carbanions or those in the parent compounds have a stronger or weaker influence on energy. This is revealed by comparing SDE-s to the H^o_{rel}-s of the parent compounds. The Relative Structure Dependent Energies (RSDE-s) are deduced by subtracting from the SDE values of radicals, cations or anions the H^o_{rel}-s of their parent compounds. A negative RSDE value means, for example, that the same structural element brings about either stronger energy decrease or a weaker energy increase in a radical, cation or anion then it causes in the parent compound.

Á. Furka, *The Structure Dependent Energy of Organic Compounds*, SpringerBriefs in Molecular Science, https://doi.org/10.1007/978-3-030-06004-6_7

7.1 Radicals

The radicals can be deduced from the methyl radical by substituting one or more hydrogen atoms with different functional groups. The SDE of radicals as well as the H^o_{rel}-s of their parent compounds are summarized in Table 7.1. The SDE of the ethyl radical is −3.9 kJ/mol. Since the H^o_{rel} of ethane is 0.0 kJ/mol the energy of the ethyl radical looks more favorable but the RSDE value is very low (−3.9 kJ/mol).

Substitutions by vinyl or phenyl groups lead to stronger effects. As indicated by H^o_{rel}-s, the presence of the double bond in propene and the benzene ring in toluene increases considerably the energy (by 82.3 and 64 kJ/mol, respectively). The same structural elements in the radicals, however, bring about only 24.3 and 12.5 kJ/mol increase, demonstrating that considering energy, the double bond or the benzene ring are less unfavorable in the radicals than in the parent compounds. The RSDE-s are considerable negative values: −58 and −51.5 kJ/mol, respectively, reflecting a special stabilization in the allyl and benzyl radicals.

The presence of ethinyl group also increases the energy in both the parent compound (H^o_{rel}: 203.8 kJ/mol) and in the radical (SDE: 161.4 kJ/mol). The energy increase, however, is lower in the case of the radical. The RSDE is a negative value: −42.4 kJ/mol, indicating a lower energy increase in the propargyl radical than in propyn.

By substitution of a hydrogen atom in methane or in the methyl radical with hydroxyl or amino group, the energies are reduced in both the parent compounds (H^o_{rel} : −11.3 and −26.6 kJ/mol, respectively) and by a higher value in the radicals (SDE −34.7 and −63.3 kJ/mol, respectively). The negative RSDE values (−23.4 and −36.8 kJ/mol) reflect that the presence of hydroxyl or amino group in the radicals is more favorable than that in the parent compound.

Table 7.1 Effect of substituents replacing one hydrogen atom in the methyl radical

Parent compound	H^o_{rel} kJ/mol	Radical	H^o_{rel} kJ/mol	SDE kJ/mol	RSDE kJ/mol
		CH_3^*	186.8	0	
$CH_3–CH_3$	0.0	$H_3C—CH_2^•$	182.9	−3.9	−3.9
$H_2C{=}CH–CH_3$	82.3	$H_2C{=}CH–CH_2^•$	211.1	24.3	−58.0
$C_6H_5–CH_3$	64.0	$C_6H_5–CH_2^•$	199.3	12.5	−51.5
$HC{\equiv}C–CH_3$	203.8	$HC{\equiv}C–CH_2^•$	348.2	161.4	−42.4
$CH_3–OH$	−11.3	$HO–CH_2^*$	152.1	−34.7	−23.4
$CH_3–NH_2$	−26.5	$H_2N–CH_2^*$	123.5	−63.3	−36.8

Radicals deduced by replacing two hydrogen atoms in the methyl radical are found in Table 7.2.

The SDE of the isopropyl radical indicates a significant energy reduction relative to the H^o_{rel} of the parent propane. The RSDE value (−30.2 kJ/mol) shows a significant stabilization in the isopropyl radical. The energy reduction effect is about the same (RSDE: −29 kJ/mol) if the methyl radical is substituted by one methyl and one hydroxyl group. Although the SDE of 1-hydroxy-ethyl radical is much lower (−55.3 kJ/mol) than that of the isopropyl radical, the RSDE values are about the

Table 7.2 Effect of substituents replacing two hydrogen atoms in the methyl radical

Parent compound	H^o_{rel} kJ/mol	Radical	H^o_{rel} kJ/mol	SDE kJ/mol	RSDE kJ/mol
$H_3C—CH_3$	1.5	$H_3C—CH^•—CH_3$	158.1	−28.7	−30.2
$CH_3–CH_2–OH$	−24.3	$H_3C—CH^•—OH$	133.5	−53.3	−29.0
CH_2F_2	−36.2	$^*CHF_2$	156.1	−30.7	5.5
CH_2ClF	−25.0	$^*CHClF$	160.6	−26.2	−1.2
CH_2Br_2	24.6	$^*CHBr_2$	205.6	18.8	−5.8
	115.2	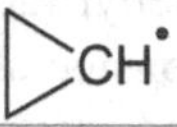 $CH^•$	320.2	133.4	16.1
	109.2	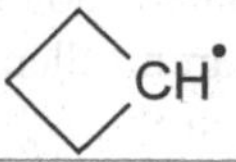 $CH^•$	291	104.2	−5.0
	0.6	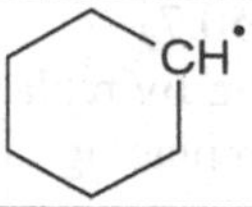 $CH^•$	177.8	−9.0	−9.6
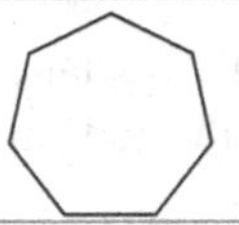	25.0	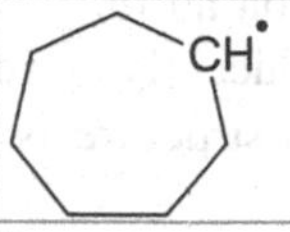 $CH^•$	200.3	13.5	−11.5
	293.5	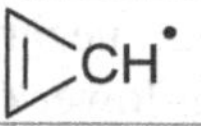 $CH^•$	483.5	296.7	3.2
	150.1	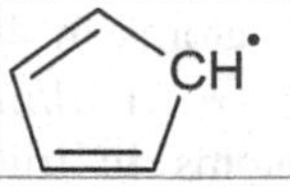 $CH^•$	257.8	71.0	−81.1
$H_2C{=}CH_2$	93.5	$H_2C{=}CH^•$	316.8	130.0	36.5
$CH_2{=}C{=}O$	62.6	$O{=}C{=}CH^*$	280.0	93.2	30.6
$CH_2{=}C{=}CH_2$	210.5	$CH_2{=}C{=}CH^*$	345.1	158.3	−52.2
HCHO	30.5	$H—C^•{=}O$	166.6	−20.2	−50.7

same. The reason is that the energy reducing effect of the hydroxyl group in the radical and in the parent compound is nearly the same.

Relative to their parents, the presence two fluorine atoms, one fluorine atom and one chlorine atom or two bromine atoms bring about only low energy changes in the substituted methyl radicals.

The presence of a radical carbon atom in cyclohexane and cycloheptane rings also causes only low relative changes (RSDE-s around −10 kJ/mol). The ring strain in cyclopropyl and cyclobutyl radicals is considerable (SDE: 133.4 and 104.2 kJ/mol, respectively) but the values are close to those appearing in their parent compounds (H^o_{rel}: 115.2 and 109.20 kJ/mol). If cyclopropenyl radical is compared to cyclopropene the situation looks similar (RSDE: 3.2 kJ/mol).

The cylopentadienyl radical is different. The consequence of the two double bonds in the ring is a 71 kJ/mol SDE. The H^o_{rel} of the parent cyclopentadiene, however, is much higher, ~150 kJ/mol. The negative RSDE value (−81.1 kJ/mol) shows that the two double bonds are "feeling better" in the radical than in the parent molecule and indicates a considerable stabilization in the cyclopentadienyl radical.

Comparing the vinyl radical to ethylene shows that the presence of the double bond in the radical causes a higher energy increase than that in the parent ethylene (RSDE: ~36 kJ/mol). The situation looks similar when the radical deduced from ketene (RSDE: ~30 kJ/mol) is considered, showing that the cumulated double bond in the radical is also more unfavorable than in the parent ketene. The opposite is observed by comparing allene with the allenyl radical. Subtracting the H^o_{rel} of allene from the SDE of the radical shows a considerable negative value for RSDE (−52.2 kJ/mol) indicating that the cumulated double bond structure is better accommodated in the radical than in the parent allene. Similarly, the SDE of the formyl radical is also significantly lower than the H^o_{rel} of the parent formaldehyde. RSDE is a considerable negative value (−50.7).

In Table 7.3 are found the radicals arising by replacing three hydrogen atoms of methyl radical by three substituents or by replacing one sp^2 or one sp carbon atom in different compounds by a radical carbon atom.

The SDE of tert-butyl radical is significantly lower than the H^o_{rel} of isobutane (RSDE: −42.6 kJ/mol) showing that the structure is more favored in the radical than in the parent isobutane.

The SDE of thrifluoromethyl radical is a high negative value (−84 kJ/mol) while the H^o_{rel} of trifluoromethane is somewhat even lower (−104 kJ/mol) indicating that considering the energy, the three fluorine atoms in the radical are somewhat less favorable than in trifluoromethane (RSDE: ~21 kJ/mol).

The presence of thee three iodine atoms in triiodomethane and in the triiodomethyl radical is unfavorable. In the latter case, however, the energy increase is somewhat lower (RSDE: −13.4 kJ/mol).

A radical carbon atom replacing one sp^2 atom in cyclopropene, benzene or anthracene rings looks unfavorable (RSDE-s are about 40–50 kJ/mol) if compared to the parent compounds. The presence of a radical sp carbon atom in acetylene, propyne or hydrogen cyanide is even more unfavorable (RSDE-s are 132.7, 134.5 and

Table 7.3 Effect of replacing three hydrogen atoms in the methyl radical

Parent compound	H^o_{rel} kJ/mol	Radical	H^o_{rel} kJ/mol	SDE kJ/mol	RSDE kJ/mol
H_3C–$CH(CH_3)_2$	−8.6	$(H_3C)_2\dot{C}$–CH_3	135.6	−51.2	−42.6
CHF_3	−104.4	F_3C^*	103.2	−83.6	20.8
CHI_3	48.4	I_3C^*	221.8	35.0	−13.4
(cyclopropene)	293.5	(cyclopropenyl radical, $C^•$)	520.7	333.9	40.4
(benzene)	76.3	(phenyl radical, $C^•$)	308.9	121.2	44.9
(anthracene)	128.5	(anthracenyl radical, $C^•$)	363.9	177.1	48.4
HC≡CH	224.5	HC≡C$^•$	544.0	357.2	132.7
H_3C–C≡CH	203.8	H_3C–C≡C$^•$	525.1	338.3	134.5
N≡CH	40.1	N≡C$^•$	327.9	141.1	101.0
CH_3CHO	0.7	H_3C–$\dot{C}$=O	135.7	−51.1	−51.8
FCHO	−53.7	F–$\dot{C}$=O	124.8	−62.0	−8.3

101 kJ/mol, respectively). In contrast, if the SDE of the acetyl radical (−51.1 kJ/mol) is compared to that of acetaldehyde (0.7 kJ/mol) it shows that the C–O double bond is more favorable in the radical than in the parent compound (RSDE −51.8 kJ/mol). A similar situation is observed if the fluoroformyl radical is compared to its parent compound but the RSDE is a small negative value (−8.3 kJ/mol).

7.2 Carbocations

The relative enthalpies of carbocations are much higher than those of the radicals for obvious reasons. The SDE values are calculated by subtracting the H^o_{rel} of methyl cation (1137.8 kJ/mol) from the H^o_{rel}-s carbocations. The carbocations deduced from the methyl cation by substituting one of its hydrogen atoms by different functional groups are listed in Table 7.4.

If the SDE value of the ethyl cation (−171.9 kJ/mol) is compared to the H^o_{rel} of its parent ethane (0.0 kJ/mol) it indicates high RSDE (−171.9 kJ/mol). This value shows that the effect of a methyl group on the energy of the cation is much stronger

Table 7.4 Substituents replacing one or more hydrogen atoms in the methyl cation

Parent compound	H^o_{rel} kJ/mol	Cation	H^o_{rel} kJ/mol	SDE kJ/mol	RSDE kJ/mol
		CH_3^+	1137.8	0	
$CH_3–CH_3$	0.0	$CH_3–CH_2^+$	965.9	−171.9	−171.9
CF_3H	−104.4	CF_3^+	977	−160.8	−56.4
$HC{\equiv}C{-}CH_3$	203.8	$HC{\equiv}C{-}CH_2^+$	1207.6	69.8	−134.0
CH_3OH	−11.3	$HOCH_2^+$	878.3	−259.5	−248.2
$H_2N–CH_3$	−26.5	$H_2N–CH_2^+$	726.9	−410.9	−384.2

than that on the energy of the ethyl radical (RSDE −3.9 kJ/mol, Table 7.1). The presence of three fluorine atoms in both the trifluoromethyl cation and its parent trifluoromethane brings about strong energy reductions (SDE: −160.8 kJ/mol and H^o_{rel} of trifluoromethane: −104.4 kJ/mol, respectively). The values indicate that considering the energy, the fluorine atoms are even more favorable in the cation than in the parent compound.

The presence of the C–C triple bond in propyne as well as in the propargyl cation increases the energy. The H^o_{rel} of propyne (~204 kJ/mol) and the much lower SDE of the cation (~70 kJ/mol) indicates that the triple bond is much less unfavorable in the cation than in the parent compound. The RSDE is significant, −134 kJ/mol.

The H^o_{rel} of methanol is −11.3 kJ/mol. The effect of the hydroxyl group on the SDE of the hydroxymethyl cation (−259.5 kJ/mol) looks a huge negative value compared to the H^o_{rel} of the parent compound, demonstrating the exceptionally strong effect of the hydroxyl group on the energy in the cation. The effect of the amino group in the aminomethyl cation looks even stronger (RSDE: −384.2 kJ/mol)!

Carbocations in which the positively charged atom is sp^2 or sp type carbon atom are listed in Table 7.5.

If the energies of the vinyl, formyl, acetyl or fluoroformyl cations are compared to those of their parent compounds high energy reductions are observed in the cations. The highest reduction (RSDE: −333.6 kJ/mol) belongs to the acetyl cation, the second highest (RSDE: −216.5 kJ/mol) to the formyl cation. The RSDE of the vinyl cation (−96.1 kJ/mol) and that of the fluoroformyl cation (−59 kJ/mol) show that a strong stabilization effect is working in the acyl cations.

The phenyl cation also shows a significant energy reduction comparing to benzene (RSDE: −105.4 kJ/mol). All the above data shows that considering the energy, the positive charge on sp^2 carbon atom is particularly favorable.

In contrast, considering their RSDE values the ethynyl, 1-propynyl and cyanide cations all show strong energy increase if their SDE-s are compared to the H^o_{rel}-s of their parent compounds (RSDE: ~305, ~181, and ~493 kJ/mol, respectively). This reflects that the triple C–C bond formed by the positively charged carbon atom is much more unfavorable than the presence of the same bond in the parent compounds.

Table 7.5 Carbocations with charged sp^2 or sp carbon atoms

Parent compound	H^o_{rel} kJ/mol	Cation	H^o_{rel} kJ/mol	SDE kJ/mol	RSDE kJ/mol
$CH_2{=}CH_2$	93.5	$CH_2{=}CH^+$	1135.4	−2.5	−96.1
H–CH=O	30.6	$H{-}C^+{=}O$	951.9	−185.9	−216.5
CH_3–CHO	0.7	$H_3C{-}C^+{=}O$	805.0	−332.8	−333.6
H–CF=O	-53.8	$O{=}C^+{-}F$	1025.0	−112.8	−59.0
benzene	76.3	phenyl cation (C^+)	1108.7	−29.1	−105.4
HC≡CH	224.5	$HC{\equiv}C^+$	1667.1	529.3	304.8
$HC{\equiv}C{-}CH_3$	203.8	$H_3C{-}C{\equiv}C^+$	1532.7	384.9	181.1
N≡CH	40.1	$N{\equiv}C^+$	1670.6	532.8	492.7

7.3 Carbanions

The structure dependent energies of carbanions are calculated by subtracting the H^o_{rel} of the methyl anion (186.7 kJ/mol) from the H^o_{rel}-s of the other carbanions. Two anions deduced from the methyl anion by substitution one of their hydrogen atom by different functional groups are listed in Table 7.6.

Both ethyl and aminomethyl anions show low energy increases when compared to their parent compounds (RSDE: ~21, ~16 kJ/mol, respectively). In this respect they differ from both radicals and carbocations.

In the trifluoromethyl anion all the three hydrogen atoms of the methyl anion are replaced by fluorine atoms. The presence of three fluorine atoms in both trifluoromethane and trifluoromethyl anion brings about strong energy reduction. This

Table 7.6 Effect of substituents replacing one or more hydrogen atoms in the methyl anion

Parent compound	H^o_{rel} kJ/mol	Anion	H^o_{rel} kJ/mol	SDE kJ/mol	RSDE kJ/mol
		CH_3^-	186.7	0	
CH_3–CH_3	0.0	CH_3–CH_2^-	207.6	20.9	20.9
H_2N–CH_3	−26.5	H_2N–CH_2^-	175.9	−10.8	15.7
CHF_3	−104.4	F_3C^-	−69.1	−255.8	−151.4

Table 7.7 Carbanions with charged sp^2 or sp carbon atoms

Parent compound	H^o_{rel} kJ/mol	Anion	H^o_{rel} kJ/mol	SDE kJ/mol	RSDE kJ/mol
$H_2C{=}CH_2$	93.5	$H_2{=}CH^-$	252.7	66.0	−27.5
HCHO	30.6	$H{-}C^-{=}O$	136.7	−50.0	−80.6
$H_3C{-}CH{=}O$	0.7	$H_3C{-}C^-{=}O$	96.0	−90.7	−91.4
$HC{\equiv}CH$	224.5	$HC{\equiv}C^-$	257.0	70.3	−154.2
$H_3C{-}C{\equiv}CH$	203.8	$H_3C{-}C{\equiv}C^-$	254.5	67.8	−136.0
$N{\equiv}CH$	40.1	$N{\equiv}C^-$	−44.8	−231.5	−271.6

reduction, however, is much stronger in the anion (RSDE: −151.4 kJ/mol) indicating that the presence of the fluorine atoms in the anion is greatly preferred.

Those carbanions in which the charged atom is sp^2 or sp carbon atom appear in Table 7.7.

The H^o_{rel} of ethylene is 93.5 kJ/mol and the SDE of the vinyl anion is only 66 kJ/mol. Tis demonstratrates that the presence of the double bond in the anion is less unfavorable than that in ethylene.

This tendency is even more emphasized in the case of formyl and acetyl anions by their RSDE values (−80.6 and −91.4 kJ/mol). The triple bonds in the ethynyl and 1-propynyl anion increases the energy in both cases (SDE: ~70 and ~68 kJ/mol, respectively). The energy increase, however, is much higher in the parent acetylene (H^o_{rel}: ~225 kJ/mol) and propyne (H^o_{rel}: ~204 kJ/mol). As a result, they show very high negative RSDE values (−154.2 and −136 kJ/mol, respectively) that seems to be in accordance with the acidity of these compounds.

An even higher negative RSDE value comes out (−271.6 kJ/mol) when the cyanide ion is compared to hydrogen cyanide. This may also indicate the acidity of hydrogen cyanide.

7.4 Homolytic Dissociation Energies and Relative Enthalpies

The experimental homolytic dissociation energies can be used to deduce from them the heats of formation of alkyl radicals and may also demonstrate their relative stability.

Table 7.8 Dissociation energies and relative stability of alkyl radicals

Parent compound	H^o_{rel} kJ/mol	D_{298} (H) kJ/mol	Radical	H^o_{rel} kJ/mol	SDE	Compound	H^o_{rel} kJ/mol	D_{298} (Br) kJ/mol
CH_4	−10.8	439.2	$CH_3^{\bullet}$	186.8	0.0	CH_3Br	8.7	301.9
CH_3–CH_3	0.0	423.3	CH_3–$CH_2^{\bullet}$	182.9	−5.9	CH_3CH_2Br	3.0	303.1
C_3H_8	1.5	412.8	H_3C—$CH^{\bullet}$(CH_3)	158.1	−30.8	$(CH_3)_2CHBr$	−9.4	309.4
H_3C—CH(CH_3)CH_3	−8.6	404.0	(H_3C)$_2$$C^{\bullet}$—$CH_3$	135.6	−53.3	$(CH_3)_3CBr$	−25.6	304.0

The data of the third column of Table 7.8 can be used to solve a discrepancy. The homolytic dissociation energies of the listed alkanes are decreasing with the increasing order of the radical carbon atom. The data of the last column, however, where the dissociation energies of alkyl bromides are summarized do not support this tendency. The data in eighth column of the table show that the relative energies of alkyl bromides themselves are also decreasing with their increasing order and this decrease (from 8.7 to −25.6 kJ/mol) is higher than the decrease of the H^o_{rel}-s of alkanes listed in the first column (from −10.8 through 0.0 to −8.6 kJ/mol). The differences in the energies of the parent compounds (that do not appear if their heats of formation are compared) influence the dissociation energies and provide the explanation of the mentioned discrepancy. The use of either H^o_{rel}-s or SDE-s of the radicals instead the dissociation energies solves the problem. Going downwards in the columns 5 and 6 the data show that the increasing relative stabilities of the radicals are properly reflected by the values of both their H^o_{rel}-s and SDE-s.

The examples mentioned above show that the use relative enthalpies in order to express the structure dependent energies of organic radicals, cations and anions is as fruitful as in the case of their parent organic compounds. The same energy determining structural factors may have stronger, weaker or even opposite effects in radicals, cations or anions than those in the parent molecules.

References

1. Argonne National Laboratory. Active thermochemical tables, version 1.112
2. Goos E, Burcat A, Ruscic B (2010) Extended third millennium ideal gas and condensed phase thermochemical database for combustion with updates from active thermochemical tables
3. Furka A (2017) Struct Chem Struct Chem 28:309–316

Chapter 8
Inorganic Compounds

8.1 Elements

In the traditional thermochemical reference system zero values are assigned to the heats of formation of the elements. Although the zero values were logical choices that work properly in practice, they tell absolutely nothing about the nature of the elements.

In the alternative reference system outlined in this book, the properties of the elements occurring in organic compounds and those of some inorganic compounds are properly reflected by their relative enthalpies.

The heats of formation and the relative enthalpies of elements are listed in Table 8.1. The zero heat of formation in the case of three elements belongs to solid state. The H^o_{rel}-s of these elements: carbon, sulfur and iodine are negative values −22.8, −41.6 and −107.6 (I_2), respectively. The standard state of bromine (Br_2) is liquid state that of the other five elements is bimolecular gaseous state. There are two high energy elements among them: fluorine (H^o_{rel}: 396.1 kJ/mol) and oxygen (H^o_{rel}: 251.7 kJ/mol). The highest H^o_{rel} belongs to ozone: 520.2 kJ/mol. The properties of these elements are in accordance with the high positive values of the H^o_{rel}-s. There are three moderately high energy elements: chlorine (H^o_{rel}: 94.2 kJ/mol), bromine (H^o_{rel} (gas state): 30.9 kJ/mol) and hydrogen (H^o_{rel}: 43.4 kJ/mol). The H^o_{rel}-s of these elements also reflect their properties. One of the gaseous elements, nitrogen has exceptionally low H^o_{rel} (−178.6 kJ/mol) and it behaves according to that. Compared to the highly reactive oxygen it shows rather an inert gas behavior.

8.2 Hydrogen Compounds of the Elements

As the data of Table 8.2 demonstrates that hydrogen forms low energy compounds with all of the elements of Table 8.1. The lowest energy compound is water and the energy gradually increases in direction of ammonia, hydrogen sulfide and methane.

Á. Furka, *The Structure Dependent Energy of Organic Compounds*, SpringerBriefs in Molecular Science, https://doi.org/10.1007/978-3-030-06004-6_8

Table 8.1 Heats of formation and relative enthalpies of elements

Compound	Composition	State	ΔH_f^o kJ/mol	H_{rel}^o kJ/mol
Carbon (graphite)	C	S	0.0	−22.8
Carbon (diamond)	C	S	1.8	−20.9
Hydrogen	H_2	G	0.0	43.4
Oxygen	O_2	G	0.0	251.7
Ozone	O_3	G	142.7	520.2
Nitrogen	N_2	G	0.0	−178.6
Sulfur	S	G	277.0	235.4
Sulfur	S	L	1.8	−39.8
Sulfur	S	S	0.0	−41.6
Fluorine	F_2	G	0.0	396.1
Chlorine	Cl_2	G	0.0	94.2
Bromine	Br_2	G	30.9	39.1
Bromine	Br_2	L	0.0	8.2
Iodine	I_2	G	62.4	−45.2
Iodine	I_2	L	13.5	−94.1
Iodine	I_2	S	0.0	−107.6

Table 8.2 Hydrogen compounds of the elements

Compound	Composition	H_{rel}^o kJ/mol	Compound	Composition	H_{rel}^o kJ/mol
Methane	CH_4	−10.8	Hydrogen fluoride	HF	−51.3
Water	H_2O	−72.5	Hydrogen chloride	HCl	−23.5
Ammonia	NH_3	−70.1	Hydrogen bromide	HBr	−10.4
Hydrogen sulfide	H_2S	−18.4	Hydrogen iodide	HI	−5.7
Hydrogen peroxide	H_2O_2	159.0	Hydrazine	N_2H_4	3.5

Among the hydrogen halides hydrogen fluoride has the lowest energy followed by hydrogen chloride, hydrogen bromide and hydrogen iodide. It is remarkable how high the difference is between the H_{rel}^o-s of water and hydrogen sulfide. H_{rel}^o of hydrogen peroxide is exceptionally high as the consequence of the unfavorable O–O bond. Similarly, the positive value of the H_{rel}^o of hydrazine also reflects the consequence of the weak N–N bond.

8.3 Carbon Sulfides and Oxides

Carbon monosulfide is a high energy compound as reflected by its H_{rel}^o in Table 8.3, while carbon monoxide is a moderately low energy compound. The difference in the properties of sulfur and oxygen is further strengthened by comparing the H_{rel}^o-s

Table 8.3 Carbon sulfides and oxides

Compound	Composition	H^o_{rel} kJ/mol	Compound	Composition	H^o_{rel} kJ/mol
Carbon monosulfide	CS	165.6	Carbon monoxide	CO	−7.5
Carbon disulfide	CS_2	11.0	Carbon dioxide	CO_2	−164.6
Carbonyl sulfide	COS	−77.0	Carbon suboxide	C_3O_2	89.6

of carbon disulfide and carbon dioxide. While carbon dioxide is a very low energy compound the value of the H^o_{rel} of carbon disulfide is positive. In carbonyl sulfide both sulfur and oxygen are present so the value of H^o_{rel} is between those of carbon disulfide and carbon dioxide. The H^o_{rel} of carbon suboxide fundamentally differs from that of carbon dioxide that is fully understandable taking into account the cumulated carbon double bond structure of the molecule.

8.4 Sulfuric Acid, Sulfur Oxides and Halides

Among the sulfur compounds listed in Table 8.4 sulfuric acid has the lowest and sulfur monoxide the highest energy.

It is interesting to note that H^o_{rel} of sulfur dioxide is lower than that of the three oxygen containing sulfur trioxide. Sulfur tetrafluoride is a moderately energetic compound. H^o_{rel} of the other three compounds of column 4 is gradually decreasing from disulfur dichloride to sulfuryl dichloride.

8.5 Nitrogen Compounds

All the nitrogen compounds listed in Table 8.5, have positive H^o_{rel}-s. The highest energy belongs to nitryl fluoride. The energy of nitrosyl halides gradually decreases from the high energy nitrosyl fluoride to the also energetic iodide.

Table 8.4 Other sulfur compounds

Compound	Composition	H^o_{rel} kJ/mol	Compound	Composition	H^o_{rel} kJ/mol
Sulfur monoxide	SO	89.2	Sulfur tetrafluoride	SF_4	22.2
Sulfur dioxide	SO_2	−86.8	Disulfur dichloride	S_2Cl_2	−8.5
Sulfur trioxide	SO_3	−59.4	Thionyl chloride	$SOCl_2$	−33.3
Sulfuric acid	H_2SO_4	−308.4	Sulfuryl dichloride	SO_2Cl_2	−53.0

Table 8.5 Nitrogen compounds

Compound	Composition	H^o_{rel} kJ/mol	Compound	Composition	H^o_{rel} kJ/mol
Nitrogen dioxide	NO_2	196.2	Nitryl chloride	$ClNO_2$	197.4
Nitrogen monoxide	NO	127	Nitrosyl fluoride	FNO	168.9
Dinitrogen monoxide	N_2O	28.8	Nitrosyl chloride	ClNO	135.4
Nitric acid	HNO_3	176	Nitrosyl bromide	BrNO	122.7
Nitryl fluoride	FNO_2	251.7	Nitrosyl iodide	INO	94.9

The nitrogen oxides of column 1 are also high energy compounds. The highest H^o_{rel} belongs to nitrogen dioxide. Comparing the H^o_{rel}-s of the three nitrogen oxides it can be seen that the more oxygens per nitrogen atoms are in the molecules the highest is the energy. Nitric acid is also a high energy compound that is particularly appreciated if the H^o_{rel} is compared to the low H^o_{rel} of sulfuric acid (−308.4 kJ/mol, Table 8.4).

References

1. Furka Á (2009) Struct Chem 20:587–604
2. Furka Á (2009) Struct Chem 20:605–616

Chapter 9
Components of the Heats of Reactions

In organic chemistry text books the feasibility of organic reactions is most often interpreted by the reactivity of the reactants. Basicity, acidity, nucleophylicity or electrophilicity are important factors that are considered. While the importance of the reactivity can't be questioned, reactivity is not the only determining factor in the reactions. There are other important factors as energy and entropy. These are the components of the driving force that determines the direction of reactions. In the case of most exothermic reactions the driving force is the heat of reaction. The role of entropy mostly becomes significant when the number of the reactant and the product molecules are not the same, or at elevated temperature [1].

Since the reactivity can usually be assigned to individual reactants their contribution to the realization of the reaction can be clearly seen. The contribution to the heat of reactions, however, can't be assigned to the reactants and products. The reason is that in the traditional thermochemical reference system the energy of the compounds is not properly reflected.

In chemical reactions compounds are formed others are disappearing. The heat of reactions depends on the energy of both the disappearing and forming compounds. The difficulty to estimate the contribution of the reactants and products to the heat of reaction is exemplified by hydrogenation of ethylene.

$$CH_2{=}CH_2 + H_2 = CH_3{-}CH_3$$

The heat of reaction is −137 kJ/mol. Since by definition the heat of formation of hydrogen is zero the heat of reaction must come from the energy difference of ethane and ethylene. It is unreasonable, however, to think that incorporation of hydrogen into ethane has no contribution to the heat of reaction.

The difficulty demonstrated above can be circumvented by the use of relative enthalpies of compounds and elements in the estimations. The examples below will show how the relative enthalpies help to identify the components of the heats of reactions.

Á. Furka, *The Structure Dependent Energy of Organic Compounds*, SpringerBriefs in Molecular Science, https://doi.org/10.1007/978-3-030-06004-6_9

9.1 The Relative Enthalpies of the Elements and Their Meaning

The relative enthalpies express the energies of organic compounds relative to properly selected reference compounds: unbranched alkanes and their derivatives, ethers, sulfides, amines and halides. The relative enthalpies do not depend on composition and there is a good reason to believe that the relative enthalpies of reactants and products can be considered as their contribution to the heats of reactions.

The meaning of the relative enthalpies of elements is demonstrated in the following two virtual reactions of formation of reference compounds, diethyl ether and ethane. The relative enthalpies of the elements are expressed in kJ/mol and rounded to one decimal number. The thermochemical data in the feasible or virtual reactions used as examples throughout this chapter refer to gas state and standard conditions.

	4 C	+	5 H_2	+ ½ O_2	=	$(C_2H_5)_2O$
H^o_{rel}	4x(–22.8)		5x43.4	251.7/2		–0.4

$$\Delta H^o_r = -0.4 - 4x(-22.8) - 5x43.4 - 251.7/2 = -252.1$$

$$\Delta H^o_f = -252.1 \text{ kJ/mol}$$

	2 C	+	3 H_2	=	C_2H_6
H^o_{rel}	2x(–22.8)		3x43.4		0.0

$$\Delta H^o_r = 0.0 - 3x43.4 - 2x(-22.8) = -84.6$$

$$\Delta H^o_f = -84.7 \text{ kJ/mol}$$

If one subtracts the sum of the H^o_{rel}-s of the participating elements—that is the H^o_{rel}-s of the reactants—from the H^o_{rel}-s of the products gets the heat of reaction in both cases. The products—diethyl ether and ethane—are reference compounds, so the heat of reaction in both cases means the heat of formation of a reference compound. This shows that the relative enthalpies of elements express their contribution to the heats of formation of reference compounds. In other words:

the relative enthalpies of the elements are the components of the heats of formation of the reference compounds.

Let's consider two example reactions in which isomer compounds are formed: dimethyl ether and ethyl alcohol.

$$2\,C + 3\,H_2 + \tfrac{1}{2}\,O_2 = CH_3OCH_3$$

$$H^o_{rel} \quad 2x(-22.8) \quad 3x43.4 \quad 251.7/2 \quad 26.4$$

$$2x(-22.8) + 3x43.4 + 251.7/2 = 210.5$$

$$\Delta H^o_r = 26.4 - 210.5 = -184.1$$

$$\Delta H^o_f = -184.1 \text{ kJ/mol}$$

$$2\,C + 3\,H_2 + \tfrac{1}{2}\,O_2 = C_2H_5OH$$

$$H^o_{rel} \quad 2x(-22.8) \quad 3x43.4 \quad 251.7/2 \quad -24.3$$

$$-2x(-22.8) + 3x43.4 + 251.7/2 = 210.5$$

$$\Delta H^o_r = -24.3 - 210.5 = -234.8$$

$$\Delta H^o_f = -234.8 \text{ kJ/mol}$$

The heats of reactions are the heats of formation of the products. The ΔH^o_r-s have four components. Three components are the same in both reactions: the contribution of the relative enthalpies of the elements. Their sum is 210.5 kJ/mol. The fourth component is different in both reactions, and express the structure dependent energy (H^o_{rel}) of the product.

The examples show that the heat of formation of a compound has two essential parts:

1. One part comes from incorporation of their constituent elements into compound.
2. The second part reflects the structure dependent relative energy of the compound.

The following reaction shows that it is useful to consider the relative enthalpies even in inorganic reactions.

$$H_2 + Cl_2 = 2\,HCl$$

$$\Delta H^o_f \quad 0.0 \quad 0.0 \quad 2(-92.3)$$

$$\Delta H^o_r = 2x(-92.3) = -184.6$$

$$H^o_{rel} \quad 43.4 \quad 94.2 \quad 2(-23.5)$$

$$43.4 + 94.2 = 137.6$$

$$\Delta H^o_r = 2x(-23.5) - 137.6 = -184.6$$

The relative enthalpies of elements express their energy relative to their incorporated state in reference organic compounds. The example shows that the H^o_{rel}-s can be used even in some inorganic reactions. The heat of reaction calculated from the ΔH^o_f-s of the reactants and product is −184.6 kJ/mol. This is equal to the heat of formation of two moles of HCl. The same result can be deduced from H^o_{rel}-s.

This calculation shows, however, that a 137.6 kJ/mol (from 43.4 and 94.2) part of the heat of reaction arises from the conversion of the elements, H_2 and Cl_2 into compound state. The rest (2x (− 23.5) = −47 kJ/mol) comes from the relative enthalpy of the two mol hydrogen chloride that reflects the structure dependent energy of the two HCl-s in the alternative thermochemical reference system.

9.2 Components of the Heats of Reactions

Applicability of relative enthalpies versus heats of formations in analysis of heats of reactions can be demonstrated by hypothetical redistribution reactions of unbranched alkanes.

	$C_{12}H_{26}$	+ C_8H_{18}	= $C_{18}H_{38}$	+ C_2H_6
H^o_{rel}:	0.0	0.0	0.0	0.0

$\Delta H^o_r = 0.0$ kJ/mol

ΔH^o_f:	−290.9	−208.4	−414.6	−84.7

$\Delta H^o_r = 0.0$ kJ/mol

The example shows that the heat of reaction calculated from both the relative enthalpies and the heats of formation is alike zero. The same is true for of all such redistribution reactions except when methane is formed. This indicates that the relative energies of all reactants and products are the same otherwise the heats of the reactions could not be the same. Based on this conclusion, rather the zero relative enthalpies can be considered as components of the heat of reaction then the unequal heats of formation.

If methane forms in a redistribution reaction the calculated heat of reaction based on ΔH^o_f-s is −10.7 kJ/mol.

	$C_{12}H_{26}$	+ C_8H_{18}	= $C_{19}H_{40}$	+ CH_4
H^o_{rel}:	0.0	0.0	0.0	−10.8

$\Delta H^o_r = -10.8$ kJ/mol

ΔH^o_f:	−290.9	−208.4	−435.1	−74.9

$\Delta H^o_r = -10.7$ kJ/mol

When the heat of reaction is calculated from both the ΔH^o_f-s *and* H^o_{rel}-s, ΔH^o_r-s are very close values: −10.7 and −10.8 kJ/mol, respectively. Considering rounding of the values and the potential experimental errors these two values are practically the same. Since the relative enthalpy of methane is −10.8 kJ/mol, this shows that the formation of methane is the only contributor to the heat of reaction that is revealed only by application of the H^o_{rel}-s in the calculations.

If a branched alkane forms in such a reaction, for example 2,2-dimethyl-propane, its relative enthalpy (−19.5 kJ/mol) also shows up as the heat of reaction.

All these support the conclusion that the relative enthalpies are good choices to express the components of the heats of reactions.

$$C_{12}H_{26} + C_8H_{18} = C_{15}H_{32} + CH_3\text{-}C(CH_3)_2\text{-}CH_3$$

H^o_{rel}:	0.0	0.0	0.0	−19.5
$\Delta H^o_r = -19.5$ kJ/mol				
ΔH^o_f:	−290.9	−208.4	−352.8	−166.0
$\Delta H^o_r = -19.5$ kJ/mol				

To strengthen the conclusion let's add a new argument by considering the esterification reaction of acetic acid. Below the formulae the heats of formation are found. The heat of reaction calculated from their values is −19.9 kJ/mol. One line below, the heats of formation are expressed as H^o_{rel}-s minus the values of the sums of the H^o_{rel}-s of the elements (in bold) that have to be added to the H^o_{rel}-s to restore from them the ΔH^o_f-s. One more row below, the sums of the added corrections to the reactants products are shown (also in bold). Their values are the same and cancel out when the heat of reaction is calculated. What remain are the relative enthalpies in the lowest row.

$$CH_3\text{-}COOH + CH_3\text{-}CH_2\text{-}OH = CH_3\text{-}COO\text{-}CH_2\text{-}CH_3 + H_2O$$

ΔH^o_f:	−432	−234.8	−444.9	−241.8
ΔH^o_f:	−139.1−**292.8**	−24.3−**210.5**	−110.7−**334.1**	−72.5−**169.2**
Korr.:		**503.3**		**503.3**
H^o_{rel}:	−139.1	−24.3	−110.7	−72.5

What determine the heat of reaction are the relative enthalpies of the reactants and products. This allows an important conclusion. In organic reactions the atoms that build up the reactants and the products do not change. What really change are the structures and the heat of reactions is determined by the structure dependent energies associated with the structures of the reactants and products.

Before proceeding further in analysis of the components of the heats of reactions let's consider three types of reactions to see the potential role of entropy in such reactions.

1. Reactions that in their chemical equations have the same number of reactant and product molecules.
2. Reactions proceeding with reduction of the number of molecules in the process
3. Reactions proceeding with increasing of the number of molecules in the process

The following examples will show that the increment of entropy (ΔS^o) have different effect on these types of reactions.

Type 1 reactions

The most obvious examples for reactions belonging to this type are substitution reactions. For the substitution reaction of chlorine with ethane, both the heat of reaction and the Gibbs energy increment $\left(\Delta G^o_r\right)$ are calculated and compared.

$$CH_3\text{-}CH_3 + Cl_2 = CH_3\text{-}CH_2\text{-}Cl + HCl$$

ΔH^o_f: −84.7 0.0 −111.7 −92.3

$\Delta H^o_r = -119.3$ kJ/mol

S^o: 0.2295 0.2229 0.2758 0.1867

$\Delta S^o = 0.0101$ kJ/mol °K
$T\Delta S^o = 3.01\ (= -0.0101 \times 298)$
$\Delta G^o_r = \Delta H^o_r - T\Delta S^o = -119.3 - 3.0 = -122.3$ kJ/mol

ΔH^o_r does not, ΔG^o_r does include the effect of entropy ($T\Delta S^o$) on the reaction. The two values are very close the difference is only ~ 3 kJ/mol. Since this is the case in other such reactions, this may allow the conclusion that in the group 1 reactions the role of entropy can practically be left out of consideration.

Type 2 reactions

The best examples for type 2 reactions are the addition reactions. A simple example is the addition of hydrogen to ethylene.

$$CH_2{=}CH_2 + H_2 = CH_3\text{-}CH_3$$

ΔH^{o}_{f}: 52.3 0.0 −84.7

ΔH^{o}_{r} = −137.0 kJ/mol

S^{o}: 0.2195 0.1306 0.2295

ΔS^{o} = −0.1206 kJ/mol °K
$T\Delta S^{o}$ = −35.9 kJ/mol (= −0.1206×298)
$\Delta G^{o}_{r} = \Delta H^{o}_{r} - T\Delta S^{o} = -137.0-(-35.9) = -101.1$ kJ/mol

The two values, ΔH^{o}_{f} and ΔG^{o}_{f} are nearly not equal, they differ by 35.9 kJ/mol. If the number of molecules is reduced in a reaction the entropy is also reduced. This is reflected by the increase of the value of ΔG^{o}_{r} relative to the value of ΔH^{o}_{r}. This is disadvantageous for these reactions and in order to proceed, the entropy increment has to be overcompensated by the heat of reaction like in this example. The heat of reaction has to have a high enough negative value to make the ΔG^{o}_{r} value also negative.

Type 3 reactions

The best examples for the reactions accompanied by increasing the number of molecules in the process are decomposition and some ring closure reactions. Since increasing the number of molecules is accompanied by increase of entropy, ΔG^{o}_{r} is expected to decrease relative to ΔH^{o}_{r}.

$$CH_2\text{-}CH_2\text{-}Cl = CH_2{=}CH_2 + HCl$$

ΔH^{o}_{f}: −111.7 52.3 −92.3

ΔH^{o}_{r} = 71.7 kJ/mol

S^{o}: 0.2758 0.2195 0.1868

ΔS^{o} = 0.1305 kJ/mol °K
$T\Delta S^{o}$ = 38.9 kJ/mol (=298×0.1305)
$\Delta G^{o}_{r} = \Delta H^{o}_{r} - T\Delta S^{o} = 71.7 - 38.9 = 32.8$ kJ/mol

As expected, the value of ΔG^{o}_{r} and ΔH^{o}_{r} differ. The difference is 38.9 kJ/mol. The ΔG^{o}_{r} is less unfavorable than ΔH^{o}_{r}.

In gas phase and at room temperature the effect of entropy in the addition and simple decomposition reactions is in the range of 35–40 kJ/mol.

The three types of reactions are demonstrated below by examples. These examples are theoretical reactions and the possibility of their experimental realization is not considered.

9.3 Reactions Preserving the Starting Number of Molecules

As seen above in the example of the type 1 reactions, when the number of product molecules is the same as the number of reactant molecules, the effect of entropy changes at room temperature can practically be left out of consideration. The main factor of the driving force of reactions is the heat of reaction. Most of these reactions are substitution procedures even when they are named otherwise. An esterification reaction, for example is also a substitution reaction since in such reaction the hydroxyl group of the acid is replaced by an alkoxy group.

Formation of alkyl halides by substitution reactions

Two reactions exemplify the halogenation reactions of alkanes. The data in the parentheses next to the formulae are relative enthalpies.

$$CH_3CH_3(0.0) + F_2(396.1) = CH_3CH_2F(-0.5) + HF(-51.3)$$

$$\Delta H_r^0 = -447.9 \text{ kJ/mol}$$

$$CH_3\text{-}\underset{\displaystyle}{\overset{\displaystyle CH_3 \atop |}{CH}}\text{-}CH_3(-8.6) + Cl_2(94.2) = CH_3\text{-}\underset{| \atop \displaystyle Cl}{\overset{\displaystyle CH_3 \atop |}{C}}\text{-}CH_3(-32) + HCl(-23.5)$$

$$\Delta H_r^0 = -141.1 \text{ kJ/mol}$$

In the first reaction the very high energy fluorine reacts with the reference compound ethane. The products are the reference compound ethyl fluoride and the low energy hydrogen fluoride. The main part of the heat of reaction comes from the incorporation of fluorine into ethyl fluoride (−396.1 kJ/mol) and formation of HF also contributes with −51.3 kJ/mol.

In the second reaction the relatively high energy chlorine reacts with the non-reference compound isobutane. The heat of reaction has three components: incorporation of chlorine into tert-butyl chloride (−94.2 kJ/mol), formation of hydrogen chloride (−23.5 kJ/mol) and transformation of isobutane into *tert*-butyl chloride (−23.4 kJ/mol).

Formation of carboxylic acid esters and amides

Formation of esters from acid and alcohol is exemplified by reaction of acetic acid and ethanol.

$$CH_3COOH(-139.1) + CH_3CH_2OH(-24.3) = CH_3COOCH_2CH_3(-110.7) + H_2O(-72.5)$$
$$\Delta H^o_r = -19.8\,kJ/mol$$

Carboxylic acids are one of the lowest energy oxygen containing carbon compounds and as a consequence, conversion of acetic acid and the moderately low energy ethyl alcohol into the also low energy ester (but less low energy than the acetic acid) is heat consuming (by 52.7 kJ/mol). Only the formation of the by product, the low energy water is that makes the whole reaction slightly exothermic.

Esters are also prepared from alcohols using acid chlorides or acid anhydrides as reagents.

$$CH_3COCl(-51.8) + CH_3CH_2OH(-24.3) = CH_3COOCH_2CH_3(-110.7) + HCl(-23.5)$$
$$\Delta H^o_r = -58.1\,kJ/mol$$

$$(CH_3CO)_2O(-156.9) + CH_3CH_2OH(-24.3) = CH_3COOCH_2CH_3(-110.7) + CH_3COOH(-139.1)$$
$$\Delta H^o_r = -68.6\,kJ/mol$$

Despite the high reactivity of the acid chlorides, as expressed by their H^o_{rel}-s they are not high energy compounds. Their reaction with the also low energy alcohol leads to the even lower energy ester so contributing to the heat of reaction with −34.6 kJ/mol. The rest of the heat of reaction (−23.5 kJ/mol), comes from formation of hydrogen chloride.

Looking to the H^o_{rel} of acetic anhydride it appears to be an even lower energy compound than the acetyl chloride. Its H^o_{rel} is even lower than that of acetic acid. Reaction of acetic anhydride with ethanol to form ester would be endothermic (by 70.5 kJ/mol). This is explained by the fact that the H^o_{rel} of the anhydride is lower than that of the product ester. In addition, the low energy alcohol is also incorporated into the ester. Only formation of the by-product acetic acid (−139.1 kJ/mol) makes the reaction exothermic.

Methyl esters can also be prepared using diazomethane as reagent. Conversion of the very low energy acid to the somewhat less low energy ester would be endothermic by 43.1 kJ/mol. The whole reaction, however, is highly exothermic $\left(\Delta H^o_r = -192.5\,kJ/mol\right)$. The energetic reagent, diazomethane is only in part responsible for that. The main contributor is the formation of the very low energy nitrogen.

$$CH_3COOH(-139.1) + CH_2N_2(57.0) = CH_3COOCH_3(-96.0) + N_2(-178.6)$$
$$\Delta H_r^o = -192.5\,\text{kJ/mol}$$

Carboxylic acid amides can be prepared from amines by acylating them with acid halides or acid anhydrides. Formation of the low energy *N*-butylacetamide from acetyl chloride and butylamine is an exothermic process (−37.2 kJ/mol). Formation of the by-product hydrogen chloride makes the reaction even more exothermic.

$$CH_3COCl(-51.8) + C_4H_9NH_2(-33.7) = CH_3CO{-}NHC_4H_9(-122.7) + HCl(-23.5)$$
$$\Delta H_r^o = -60.7\,\text{kJ/mol}$$

The role of the carboxylic acid anhydrides in acylation of amines is similar to that of acylation of alcohols. Formation of the amide from the two reactants is endothermic by 67.9 kJ/mol. Only the formation of the low energy acetic acid makes the heat of reaction negative.

$$(CH_3CO)_2O(-156.9) + C_4H_9NH_2(-33.7) = CH_3CO{-}NHC_4H_9(-122.7) + CH_3COOH(-139.1)$$
$$\Delta H_r^o = -71.2\,\text{kJ/mol}$$

Esters can be prepared from carboxylic acids using phosgene, too. One example is preparation of ethyl chloroformate that is a low energy compound. Its formation from the two reactants is in itself exothermic (by −72.6 kJ/mol). Formation of the by-product hydrogen chloride is an extra contribution to the heat of reaction.

$$COCl_2(-21.6) + C_2H_5OH(-24.3) = Cl{-}CO{-}OC_2H_5(-120.5) + HCl(-23.5)$$
$$\Delta H_r^o = -96.1\,\text{kJ/mol}$$

Phosgene and ethanol may form a different ester, diethyl carbonate, too. Like in the previous reaction, formation of the very low energy product is in itself exothermic (by −86.7 kJ/mol). Formation of the two hydrogen chloride molecules makes the process even more exothermic.

$$COCl_2(-21.6) + 2\,C_2H_5OH(-24.3) = (C_2H_5O)_2CO(-156.9) + 2HCl(-23.5)$$
$$\Delta H_r^o = -133.7\,\text{kJ/mol}$$

Formation of carboxylic acid anhydrides

Formation of acid anhydrides is demonstrated by one example. Conversion of acetic acid and acetyl chloride into acetic anhydride is endothermic by 34 kJ/mol. Only formation of the low energy by-product hydrogen chloride makes the process less endothermic.

$$CH_3COOH(-139.1) + CH_3COCl(-51.8) = (CH_3CO)_2O(-156.9) + HCl(-23.5)$$
$$\Delta H_r^o = 10.5\,\text{kJ/mol}$$

9.4 Reactions Proceeding with Reduction of the Number of Molecules

The chemical procedures in which the number of product molecules is reduced relative to the number of reactant molecules are best exemplified by addition reactions. The consequence of the reduction of the number of molecules is the reduction of entropy. The reaction proceeds only if the released heat is large enough to compensate for the reduction of entropy. Some cyclization reactions are also followed by entropy decrease.

Additions to unsaturated hydrocarbons

One of the typical addition reactions is the hydrogenation of alkenes. The heat of reaction is large enough to compensate for reduction of entropy. As mentioned the $T\Delta S$ component in such reactions is around −30 to −40 kJ/mol. The value of ΔH_r^o of the hydrogenation reaction of ethylene is −136.9 kJ/mol. One of the components of ΔH_r^o is incorporation of hydrogen (−43.4 kJ/mol) and the other comes from conversion of the unsaturated ethylene into the saturated ethane (−93.5 kJ/mol). The high heat of reaction overcompensates the effect of reduction of entropy.

$$CH_2{=}CH_2(93.5) + H_2(43.4) = CH_3{-}CH_3(0.0)$$
$$\Delta H_r^o = -136.9\,\text{kJ/mol}$$

One may question whether the 93.5 kJ/mol is really expressing the energy increase of ethylene caused by unsaturation. In the following virtual reaction the unsaturated ethylene is converted to the saturated butane by addition of the saturated ethane. The absolute values of the heat of reaction and the H_{rel}^o of ethylene are practically the same. The heat of this reaction demonstrates that the relative enthalpy of ethylene reflects the energy of its unsaturation.

$$CH_2{=}CH_2(93.5) + CH_3{-}CH_3(0.0) = CH_3CH_2CH_2CH_3(-0.2)$$
$$\Delta H_r^o = -93.7\,\mathrm{kJ/mol}$$

The relative enthalpy of chlorine is higher than that of hydrogen. Its incorporation into 1,2-dichloro-ethane and saturation of ethylene in the same process results in a large heat of reaction. Almost all of the high energy of ethylene plus the also high energy of chlorine are converted into the heat of reaction.

$$CH_2{=}CH_2(93.5) + Cl_2(94.2) = Cl{-}CH_2{-}CH_2{-}Cl(5.8)$$
$$\Delta H_r^o = -181.9\,\mathrm{kJ/mol}$$

The relative enthalpy of gas phase bromine is lower (39.1 kJ/mol) than that of chlorine (94.2 kmol), consequently the heat of reaction of its addition is also lower (−122.1 kJ/mol).

$$CH_2{=}CH_2(93.5) + Br_2(39.1) = Br{-}CH_2{-}CH_2{-}Br(10.5)$$
$$\Delta H_r^o = -122.1\,\mathrm{kJ/mol}$$

The following reaction exemplifies addition of a compound, hydrogen chloride replacing the elemental chlorine. Saturation of ethylene contributes to the heat of reaction as above by 93.5 kJ/mol, but replacement of the high energy chlorine (94.2 kJ/mol) by the low energy hydrogen chloride (−23.5 kJ/mol) reduces the heat of reaction from −181.9 kJ/mol to −71.6 kJ/mol.

$$CH_2{=}CH_2(93.5) + HCl(-23.5) = CH_3{-}CH_2{-}Cl(-1.6)$$
$$\Delta H_r^o = -71.6\,\mathrm{kJ/mol}$$

Replacement of HCl by water brings about a further reduction in ΔH_r^o. Since the relative enthalpy of water is much lower (−72.5 kJ/mol) than that of hydrogen chloride, despite of formation of the low energy ethanol (−72.5 kJ/mol) the heat of the reaction is only −45.3 kJ/mol. This is close to just compensate the entropy reduction.

$$CH_2{=}CH_2(93.5) + H{-}O{-}H(-72.5) = CH_3{-}CH_2{-}OH(-24.3)$$
$$\Delta H_r^o = -45.3\,\mathrm{kJ/mol}$$

If hydrogen is added to acetylene to form ethylene ΔH_r^o is a higher negative value than in saturation of ethylene (−136.9 kJ/mol). The reason is that the relative energy difference of acetylene and ethylene is higher than that of ethylene and ethane.

$$HC{\equiv}CH(224.5) + H_2(43.4) = CH_2{=}CH_2(93.5)$$
$$\Delta H_r^o = -174.4\,\mathrm{kJ/mol}$$

Benzene can be saturated to cyclohexane by addition of hydrogen. From the considerable heat of reaction (−205.9 kJ/mol), −75.7 kJ/mol comes from the energy difference of benzene and cyclohexane. The rest, −130.2 kJ/mol, represents the heat evolved by incorporation of the three hydrogen molecules. Although in this reaction a considerable entropy reduction is expected its effect is overcompensated by the high heat of reaction.

	benzene	+ $3H_2$	= cyclohexane
H_{rel}^o	76.3	43.4	0.6

$$\Delta H_r^o = -205.9\ \mathrm{kJ/mol}$$

The relative enthalpy of benzene reflects its energy relative to the saturated reference alkanes and there is no need to use hydrogenation in order to estimate its energy or to compare the energy to a non-existent cyclohexatriene. It is more demonstrative to show that the relative enthalpy of the triply unsaturated benzene is lower than the H_{rel}^o of the single unsaturated ethylene (93.5 kJ/mol).

Additions to functionalized derivatives of hydrocarbons

The relative enthalpy of alkyl nitriles is close to zero and formation relatively low energy alkyl amines in the below reduction reaction is in itself a heat evolving process (−34.7 kJ/mol in the example reaction). The rest of the considerable heat of reaction is the result of the incorporation of the two hydrogen molecules.

$$CH_3CH_2{-}CN(1.4) + 2H_2(43.4) = CH_3CH_2CH_2{-}NH_2(-34.7)$$
$$\Delta H_r^o = -122.9\,\mathrm{kJ/mol}$$

Ethyl alcohol can be prepared from acetaldehyde by addition of hydrogen. Formation of the relatively low energy alcohol from the almost zero energy acetaldehyde contributes to the heat of reaction with −25 kJ/mol. The rest evolves as a result of incorporation of hydrogen.

$$CH_3{-}CH{=}O(0.7) + H_2(43.4) = CH_3{-}CH_2{-}OH(-24.3)$$
$$\Delta H_r^o = -68.4\,\mathrm{kJ/mol}$$

If ethane (0.0 kJ/mol) and water (−72.5 kJ/mol) forms in hydrogenation, the main components of the heat of reaction (−160 kJ/mol) are the contributions of incorporation of the two molecules of hydrogens (−86.8 kJ/mol) and formation of water (−72.5 kJ/mol).

$$CH_3{-}CH{=}O(0.7) + 2H_2(43.4) = CH_3{-}CH_3(0.0) + H_2O(-72.5)$$
$$\Delta H_r^o = -160\,\text{kJ/mol}$$

The ring opening reaction of ethylene oxide by addition of hydrogen, leads to formation of ethanol. In the conversion of the high energy ethylene oxide to the relatively low energy ethanol 146.9 kJ/mol evolves. The rest of the heat of reaction comes from incorporation of hydrogen.

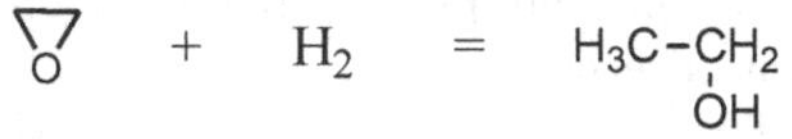

	ethylene oxide	+ H_2	= $H_3C{-}CH_2OH$
H^o_{rel}	122.6	43.4	–24.3

$$\Delta H^o_r = -190.3\ \text{kJ/mol}$$

As shown below ΔH_r^o is reduced if hydrogen is replaced in ring opening with the low energy water despite the formation of ethylene glycol that has lower energy than ethanol.

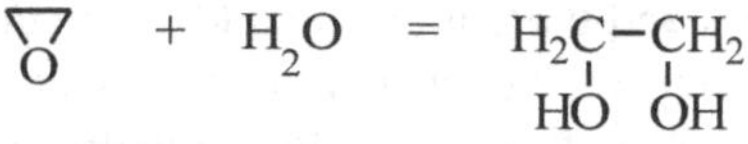

	ethylene oxide	+ H_2O	= $H_2C(OH){-}CH_2(OH)$
H^o_{rel}	122.6	–72.5	–52.9

$$\Delta H^o_r = -103\ \text{kJ/mol}$$

Formation of acetals

It is well known that carbohydrates form cyclic hemiacetals. Cyclization occurs despite the unfavorable decrease of entropy. This is the reason why the acetal forming reactions are analyzed in some details.

In formation of non-cyclic acetals two product molecules are formed from three reactant molecules. For this reason the entropy is reduced. The entropy decrease needs to be overcompensated by the heat of reaction. In the following example the acetal forming reaction is exothermic. The heat of reaction comes from two sources. First, the H^o_{rel} of the forming 1,1-diethoxy ethane is lower by 35.4 kJ/mol than that of the reactant acetaldehyde. Second, the relative enthalpy of the forming water is lower than the sum of the relative enthalpies of the two reactant ethanol molecules by 23.9 kJ/mol.

$$CH_3CHO(0.7) + 2\,CH_3CH_2OH(-24.3) = CH_3CH(OCH_2CH_3)_2(-34.7) + H_2O(-72.5)$$
$$\Delta H_r^o = -59.3\,\text{kJ/mol}$$

The relatively low energy of acetals if compared to aldehydes is the result of bonding of two oxygen atoms to the same carbon atom that is energetically more favorable than double bonding of a single oxygen atom. This makes understandable why the γ and δ hydroxy aldehydes as well as the carbohydrates tend to form hemiacetals.

The hemiacetals are non-stable compounds and, as a consequence, their heats of formation are not available. The relative enthalpy of the hemiacetal that forms in the following reaction is estimated by taking into account that it has a free hydroxyl group that reduces the energy. The relative enthalpy of acetaldehyde diethhylacetal (−34.7) was corrected by −24.3 kJ/mol that is the contribution of the hydroxyl group to the relative enthalpy of ethyl alcohol. The calculated heat of the hemiacetal forming reaction just compensates for the entropy reduction. This makes probable that the reaction leads to equilibrium.

$$CH_3CHO + CH_3CH_2OH = CH_3CH(OH)(OCH_2CH_3)$$

H^o_{rel} 0.7 –24.3 (–59)

$\Delta H^o_r = -35.4$ kJ/mol

In the case of formation of cyclic hemiacetals the number of molecules is not reduced.

The entropy decreases because of formation of the cyclic hemiacetal but by a somewhat smaller degree compared to a reaction accompanied by reduction of the number of molecules. For this reason formation of the cyclic hemiacetal is more favored than that of a non-cyclic one.

When the alcohols are replaced by diols in the acetal synthesis, cyclic acetal forms. Like above, the number of rectant and product molecules is the same. For this reason formation of the product is strongly favored.

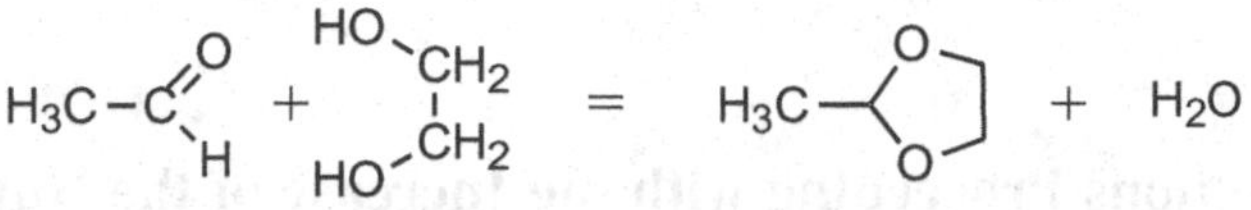

$$H_3C{-}CHO + HO{-}CH_2{-}CH_2{-}OH = H_3C{-}\text{(1,3-dioxolane)} + H_2O$$

H^o_{rel} 0.7 –52.9 –15.8 –72.5

$H^o_r = -36.1$ kJ/mol

As exemplified by the following reaction the heat of acetal forming reaction of formaldehyde is more favorable than that of acetaldehyde $(\Delta H^o_r - 59.3\,\text{kJ/mol})$ by about 10 kJ/mol. This shows, that the acetal formation from formaldehyde is more favored than that from acetaldehyde.

$$HCHO(30.6) + 2CH_3CH_2OH(-24.3) = HCH(OCH_2CH_3)_2(-14.9) + H_2O(-72.5)$$
$$\Delta H_r^o = -69.4\,kJ/mol$$

According to rough estimations hydration of formaldehyde is also favored by −10 kJ/mol relative to that of acetaldehyde.

It is well known that acetaldehyde tends to form paraldehyde that is also an acetal type compound and for this reason its energy is low. For this reason the heat of reaction is high and well over-compensates the effect of entropy reduction (by more than 60 kJ/mol).

3 $H_3C-C(=O)H$ = (2,4,6-trimethyl-1,3,5-trioxane; H_3C, CH_3, CH_3)

H^o_{rel} 0.7 –143.9.0

$$\Delta H_r^o = -146\ kJ/mol$$

The following example shows that the heat of reaction of thioacetal formation is higher than that of acetal formation. The main reason is not the formation of a lower energy thioacetal since its H^o_{rel} is close to zero, but that of the low energy water well overcompensates the disappearance of the two molecules of almost zero energy thiols.

	$HCHO$ +	2 CH_3CH_2SH =	$HCH(SCH_2CH_3)_2$ +	H_2O
H^o_{rel}	30.6	(–3)	–1.9	–72.5

$$\Delta H_r^o = -99\ kJ/mol$$

9.5 Reactions Proceeding with the Increase of the Number of Molecules

The chemical reactions in which the number of product molecules is increased relative to the number of reactant molecules are accompanied by increase of entropy. This entropy increase contributes to the driving force of the processes.

Formation of carboxylic acid chlorides

Formation of acetyl chloride from acetic acid and thionyl chloride is endothermic by 120.6 kJ/mol. The low energy by-products, hydrogen chloride and sulfur dioxide, however, makes the process less endothermic.

$$CH_3COOH(-139.1) + SOCl_2(-33.3) = CH_3COCl(-51.8) + HCl(-23.5) + SO_2(-86.8)$$
$$\Delta H_r^o = 10.3\,\mathrm{kJ/mol}$$

In this process three product molecules are formed from two reactants. As a consequence, the entropy increase contributes to the driving force of the reaction.

The hydroperoxide rearrangement

In the hydroperoxide rearrangement a single cumene hydroperoxide molecule rearranges to two product molecules: phenol and acetone.

H^o_{rel}	240.5	22.8	–29.9

$$\Delta H_r^o = -247.6\ \mathrm{kJ/mol}$$

Since the cumene hydroperoxide is a high energy compound ($H^o_{rel} = 240.5\,\mathrm{kJ/mole}$), the main component of the driving force of reaction is its transformation to the relatively low energy phenol. Formation of the low energy acetone also contributes as well as the increase of entropy, that is the consequence of the increase of the number of molecules in the process.

Cyclysation reactions

Cyclisation itself is accompanied by an entropy decrease. This is demonstrated by comparing the entropies of three cycloalkanes to those of their 1-alkene isomers.

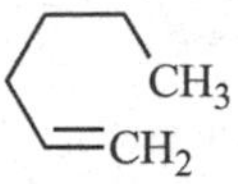

S^o: 0.384 0.298

$\Delta S^o = -0.086$ kJ/mol.°K
$T\Delta S^o = -25.6$ kJ/mol

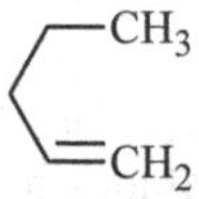

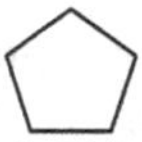

S^o: 0.305 0.293

$\Delta S^o = -0.012$ kJ/mol.°K
$T\Delta S^o = -3.5$ kJ/mol

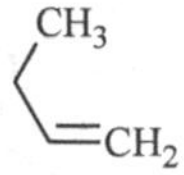

S^o: 0.267 0.265

$\Delta S^o = -0.002$ kJ/mol.°K
$T\Delta S^o = -0.6$ kJ/mol

It can be seen that these hypothetical cyclizations are really accompanied by entropy decrease. The magnitudes of the entropy decreases, however, are smaller than the entropy increments associated with the reactions proceeding with increasing of the number of molecules in the process. The $T\Delta S^o$ values of the reactions where the number of the product molecules are higher than those of the reactant ones are in the range of 35–40 kJ/mol. As a consequence, the ring closure reactions accompanied by increase of the number of molecules are thermodynamically favored.

Cyclization reactions accompanied by increase of the number of molecules are exemplified by cyclization of diols to cyclic ethers and cyclization of hydroxy acids to lactones.

Cyclization of the low energy 1,5-pentanediol to the almost zero energy tetrahydropyran is endothermic by 48.9 kJ/mol. Formation of the low energy by-product water, however, makes the process moderately exothermic. The entropy also increases because of formation of two molecules from one so the driving force of the reaction is further increased.

$HO\text{-}(CH_2)_5\text{-}OH$ = [tetrahydropyran] + H_2O

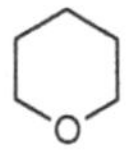

H^o_{rel} −43.8 5.1 −72.5

$\Delta H^o_r = -23.6$ kJ/mol

The analogous cyclization of the reactant 1,4-butanediol is thermoneutral ($\Delta H_r^o = 0.0\,\mathrm{kJ/mol}$). The reaction is favored, however, by the entropy increase.

$$\mathrm{HO\text{-}(CH_2)_4\text{-}OH} = \text{(tetrahydrofuran)} + H_2O$$

H^o_{rel} −48.4 24.1 −72.5

$\Delta H_r^o = 0.0$ kJ/mol

The hydroxy acids may form cyclic lactones by ring closure. The cyclization is accompanied by formation of a second molecule: water. The following example shows that the estimated relative enthalpy of γ-hydroxybutyric acid is much lower than that of the forming cyclic ester. Even the formation of the low energy by-product water does not make the process exothermic. Because of entropy increase, however, the cyclization is favored particularly at elevated temperature.

$$\text{HO–CH}_2\text{CH}_2\text{CH}_2\text{–C(=O)OH} = \text{(γ-butyrolactone)} + H_2O$$

H^o_{rel} (−161) −73.3 −72.5

$\Delta H_r^o = 15.2$ kJ/mol

Other reactions proceeding with the increase of the number of molecules

There are other reactions than cyclization that proceed with the increase of the number of molecules. One example is preparation of alkyl halides from alcohol with thionyl chloride.

$$\begin{aligned} &CH_3CH_2OH(-24.3) + SOCl_2(-33.3) = CH_3CH_2Cl(-1.6) + SO_2(-86.8) \\ &\quad + HCl(-23.5) \\ &\Delta H_r^o = -54.3\,\mathrm{kJ/mol} \end{aligned}$$

Ethanol is a relative low energy compound. Its transformation to the zero energy ethyl chloride is endothermic by 22.7 kJ/mol. Conversion of thionyl chloride into sulfur dioxide and hydrogen chloride is exothermic by −77 kJ/mol. This makes the whole process exothermic. In addition, the entropy is also increased because of formation of three molecules from two. This adds to the driving force of reaction.

A second example is decarboxylation of acetic acid. This is a hypothetic reaction that does not realize at room temperature.

$$\begin{aligned} &CH_3COOH(-139.1) = CH_3CH_3(0.0) + CO_2(-164.6) \\ &\Delta H_r^o = -25.5\,\mathrm{kJ/mol} \end{aligned}$$

Because of the very low relative enthalpy of carbon dioxide the decarboxylation would be an exothermic reaction. In addition entropy increases because two molecules forms from one and this adds to the driving force. The monocarboxylic acids, including the acetic acid are stable compounds and do not decompose at room temperature. This example shows that there are thermodynamically favored reactions that do not proceed. The reason is the high activation energy.

The examples described above represent an attempt to deduce the heats of the organic reactions from the structure dependent energies of the reactants and products. This is possible because the kind and number of elements that build up the reactants are the same as those that are present in the products. For this reason the energy associated with them cancel out. So the contribution of the energies of the constituent elements to the heats of reactions can be neglected. What really change during a chemical reaction are the structures. The analysis of the dependence of the heats of reactions on structures changes and the energies associated with them, offers a possibility to deeper understanding the nature of organic reactions.

Reference

1. Furka Á (2009) Struct Chem 20:781–788

Printed in the United States
By Bookmasters